Project cost estimating

ENGINEERING MANAGEMENT

Other titles in the series

Civil engineering insurance and bonding, P. Madge
Marketing of engineering services, J. B. H. Scanlon
Construction planning, R. H. Neale and D. E. Neale
Civil engineering contracts, S. H. Wearne
Managing people, A. S. Martin and F. Grover (editors)
Control of engineering projects, S. H. Wearne (editor)
Project evaluation, R. K. Corrie (editor)
Management of design offices, P. A. Rutter and A. S. Martin (editors)
Financial control, N. M. L. Barnes (editor)
Principles of engineering organization, S. H. Wearne
Construction management in developing countries, R. K. Loraine

ENGINEERING MANAGEMENT

Project cost estimating

Edited by Nigel J. Smith

Thomas Telford, London

Published by Thomas Telford Ltd, Thomas Telford House, 1 Heron Quay, London E14 4JD

Distributors for Thomas Telford books are
USA: American Society of Civil Engineers, Publications Sales Department, 345 East 47th Street, New York, NY 10017-2398
Japan: Maruzen Co. Ltd, Book Department, 3-10 Nihonbashi 2-chome, Chuo-ku, Tokyo 103
Australia: DA Books and Journals, 11 Station Street, Mitcham 2131, Victoria

First published 1995

British Library Cataloguing in Publication Data
Estimating
1. Engineering industries. Estimating
I. Smith, Nigel J. II. Series
620.

ISBN 0 7277 2032 5

Typeset in Great Britain by MHL Typesetting Ltd, Coventry
Printed and bound in Great Britain by Redwood Books, Trowbridge, Wiltshire

Preface

The objective of an estimate for a project is to provide the most realistic prediction possible of cost and time, no matter at what stage the estimate is undertaken.

Estimates can be prepared at any stage during a project, but in order to make use of an estimate it is important to consider its purpose, the stage of the project when it was prepared and the degree of risk inherent in the project. To have any meaning the purpose of any estimate must be linked to the stage of the project and to the data available.

Engineers have become all too familiar with the sequence of unrealistic estimates which are produced and then superseded during the life of a project. In many cases there is insufficient investigation of the reasons for the poor quality of the estimate and consequently estimates have not always been treated seriously. This trend has been reinforced by a common tendency to regard estimating as a separate function which should be isolated from the engineering and management of a project.

As one of the series of engineering management guides, the primary aim of this book is to offer practical advice and further information on the understanding, preparation and use of estimates in the civil engineering industry. The guide is split into two parts; the first deals with estimating at different stages of construction projects, and the second with the practice of estimating, in terms of the human characteristics of the estimator and the use of computer-based systems.

Acknowledgements

I am particularly grateful to the contributors to this guide and to Stephen Wearne who provided valuable assistance and guidance throughout its preparation.

I would also like to thank many civil engineers and other professionals in the construction industry for their support and assistance with the development of this guide. I would particularly like to thank the staff of the Project Management Group at the University of Manchester Institute of Science and Technology and a number of individuals including E. P. Delany.

The information on which the case study in chapter 7 is based is reproduced by permission of Severn Trent Water Ltd.

Finally, I would like to thank Joan Carey for processing and checking each of the many draft versions of this book.

N. J. Smith

Contents

Part 1. Estimating for construction projects

This part examines estimating techniques and practice at a number of different stages in a project. The early estimate is particularly important as it influences the client's brief and can determine the viability of the entire project. Despite the lack of accurate data and the presence of unquantifiable factors, it is therefore important to make the first estimate realistic.

The estimating processes most familiar to civil engineers, design estimates and construction estimates are covered. Estimating at the design stage is not always considered as seriously as it should be. This guide concentrates on data capture, estimating for budget and time control and the value of feedback into design management. Estimating the cost of construction is what most civil engineers think of when estimating is mentioned.

Estimating for process plants involves different disciplines and so estimating construction costs is specialized in nature. Particular attention is given to fixed-price estimating and the greater emphasis on technical and commercial estimating in this sector.

The later stages of a project, the role of the estimate in tender evaluation and the use of project control reports are also reviewed.

This part concludes with a detailed cost estimating case study of a water treatment plant from its conceptual design through detailed design to project audit.

1 Purposes and stages of estimating

This chapter provides an overview of the purpose and nature of estimating for civil engineering works.

In all estimates we are trying to make predictions, and the further into the future we go, the more uncertain and more complex the process becomes. If we wish to make progress we have to make decisions—in this case, decisions concerning the type of estimating technique and the most appropriate stage of the project at which to undertake an estimate.

The reason for estimating is to provide the most realistic prediction possible of time and cost at any given stage in a project. Over the life of a project the estimator should be able to produce a series of estimates, from an early estimate at inception to the final account, in which the increasing accuracy of the estimate is reflected in the decreasing extent of risk and uncertainty. All estimates take the form of base estimates, plus allowances for uncertainties and specific contingencies as required.

Estimating at project stages

In theory, the unknowns (i.e. the risk associated with the project) can be shown to decrease as the project progresses until at the project completion the final cost is known with certainty. In practice, this situation corresponds with a better understanding of project needs which often leads to increases in the project cost. The question of accuracy is often raised. A perfect estimate is not possible; the best possible estimate will always contain a number of key risks. The goal of the estimator is a practicable level of accuracy.

It is important to consider the project stages at which estimates can realistically and usefully be produced so that there is a sound basis for deciding whether or not to proceed to the next stage and so that there is a basis on which to manage that stage. This management activity should include cost reviews, risk assessments and the monitoring of progress and expenditure.

The number of stages in a project is influenced by the procurement strategy adopted. Traditional civil engineering projects can be divided into six stages

- identification
- appraisal
- definition
- approval
- implementation
- operation.

These stages may not be appropriate for every project and cannot be adhered to exclusively, but they do offer a rational and structured approach which is applicable to many construction projects. Fig. 1 shows the sequence of these project stages and indicates the types of estimate used in each stage. At the implementation stage there can be two types of estimate—pre-tender and post-contract—which take into account the changes of scope during the tendering and award processes.

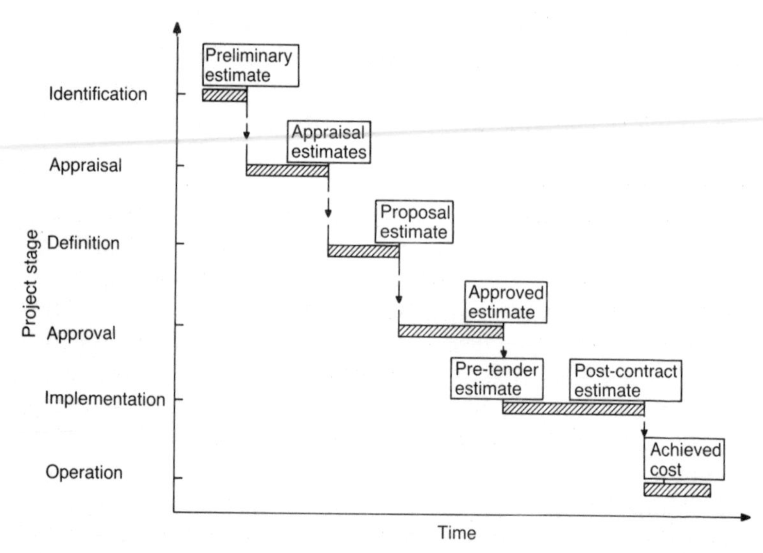

Fig. 1. Variation in type of estimate with changes in project stages

- *Preliminary estimate*: an initial estimate at the earliest possible stages. It is likely that no design data will be available and that there will be only a crude indication of the project size or capacity. The preliminary estimate is likely to be of use in the provisional planning of capital expenditure programmes.
- *Appraisal estimates*: sometimes known as feasibility estimates. These are directly comparable estimates of the alternative schemes under consideration. An appraisal estimate should include all costs which will be charged against the project in order to provide the best estimate of anticipated total cost. If it is to be used to update the initial figure in the forward budget, then it must be escalated to a cash estimate.
- *Proposal estimate*: an estimate for the selected scheme. A proposal estimate is usually based on a conceptual design and design study specifications and produced as a cash estimate to support the case for the development of an outline design.
- *Approved estimate*: a modified version of the proposal estimate to reflect the client's views, which is intended to provide the basis for project cost control.
- *Pre-tender estimate*: a refinement of the approved estimate based on the definitive design work using the information provided in the tender documents which should be used during bid evaluation as a marker against which bids can be assessed. At this stage this estimate must reflect the total capital cost of the project and all charges and fees and allowance or contingency for the client's risk.
- *Post-contract estimate*: a further refinement to reflect the prices in the contract awarded. This requires redistribution of the money in, for example, the bill of quantities in an admeasurement contract and assists more efficient management of the contract.
- *Achieved cost*: a record of the actual costs of the job in order to review performance and provide data for future projects. It is useful to compare the actual use and expenditure of allowances and contingencies with those included in the various estimates.

Table 1. Information available to estimator at each project stage

Project stage	Identification	Appraisal	Definition
Activities	Indentification of project by promoter, possibly with consultant	Appraisal of the identified project by promoter and/or consultant, including basic requirements, alternative schemes and recommendation of preferred scheme Normally requires funding of a feasibility study by a consultant	Definition of preferred scheme including basic design data, conceptual design, technical specifications, consultant's programme to completion, construction appraisal, contract strategies and estimate of final cash cost Requires funding of a design study by a consultant
Product	Inclusion in forward budget	Choice of preferred scheme	Project definition report for use by promoter in a submission for sanction
Available information for estimate	No design Capacity/size/ location only	Preliminary designs of alternatives	Conceptual design
Type of estimate required	Preliminary	Appraisal	Proposal
Probable estimator	Promoter with consultant (using historical cost database)	Feasibility study consultant	Design study consultant

Table 1. (contd)

Approval	Implementation	Operation
Consideration of submission by promoter	Implementation of approved project including detailed design, issue of tender enquiries, assessment of tenders, placing of contracts, construction, completion and commissioning	Operation of new asset by promoter Evaluation of project by promoter and consultant
Sanction for defined project Basis for cost control	Basis for assessment of tenders and ongoing monitoring of costs and progress against approved estimate	Historical cost and productivity database
Conceptual design	Tender documents	Completed contract
Approved (i.e. proposal with modifications)	Pre-tender/post-contract (refinement of approved)	Achieved costs
Consultant/promoter	Consultant	Consultant/promoter

Table 2. Main estimating techniques and their data requirements

Data requirements	Estimating technique				
	Global	Factorial	Man-hours	Unit rate	Operational
Project data	Size/capacity Location Completion date	List of main installed plant items Location Completion date	Quantities Location Key dates Simple method statement Completion date	Bill of quantities (at least main items) Location Completion date	Materials quantities Method statement Programme Completion date
Basic estimating data: potential problems, risks, uncertainties and peculiarities of the project	Achieved overall costs of similar projects (adequately defined) Inflation indices Market trends General inflation forecasts	Established factorial estimating system Recent quotes for main plant items Inflation indices (for historical prices) Market trends General inflation forecasts	Hourly rates Productivities Overheads Materials costs Hourly rate forecasts Materials costs forecasts Plant data	Historical unit rates for similar work items Preliminaries Inflation indices Market trends General inflation forecasts	Labour rates and productivities Plant costs and productivities Materials costs Overhead costs Labour rate forecasts Materials costs forecasts Plant capital and operating costs forecasts

Table 1 summarizes the activities comprising each of the project stages and the information available to the estimator. The availability of data is a significant factor in the estimator's selection of the most appropriate estimating technique.

Types of estimating technique

The five most common estimating techniques are classified, as shown in Table 2, in terms of their project data requirements and estimating data requirements. These techniques are described in detail in other chapters of this guide, but they may be summarized as follows. The first four make use of historical data and require that historic rates are selected only from an adequate sample of similar work in similar locations using similar resources and technology—this is not always possible.

- *Global*: a crude estimating technique which relies on the existence of data for similar projects assessed purely on a single characteristic such as size, capacity or output.
- *Factorial*: a technique used widely on process plants where the key components can be easily identified and priced and all other works are calculated as factors of these components.
- *Man-hours*: the technique most suitable for labour-intensive operations, like maintenance or mechanical erection; work is estimated in total man-hours and costed in conjunction with plant and material costs.
- *Unit rate*: a technique based on the traditional bill of quantities approach where the quantities of work are defined and measured in accordance with a standard method of measurement (in the UK this is usually CESMM3* or the *Method of measurement for highways*†).
- *Operational*: a complex procedure of considering the constituent operations necessary to construct the works and estimating the labour, plant and materials costs together with the overheads.

---

* Institution of Civil Engineers. *Civil engineering standard method of measurement*, 3rd edn. Thomas Telford, London, 1991.

† Department of Transport. *Method of measurement for highways*. HMSO, London, 1987.

Estimate summaries

One of the greatest problems facing an estimator is the lack of continuity of data, methods or personnel, throughout the life of a project. This is largely due to the long gestation periods of many projects and the involvement from time to time of differing parties, including consultants, suppliers and contractors. One method of alleviating some of these difficulties is to use a standard summary form.

The aim of an estimate summary is to facilitate the production of estimates which are directly comparable and could form a complete cost history of the project. It also has the advantage that it should force estimators to consider thoroughly factors which are often given insufficient attention

- contingencies and tolerances, and hence risk and uncertainty
- cash flow based on a realistic programme
- inflation and currency fluctuations, where applicable.

The use of summaries should avoid confusion, aid comparison between different estimates for the same project and avoid omissions. It is essential that all estimates are accompanied by

- a clear description of the project, including outline specifications and drawings of the works
- a realistic programme of works
- a list of all exclusions from the estimate and/or programme and a list of all key assumptions.

Bibliography

University of Manchester Institute of Science and Technology. *Guide to cost estimating for overseas construction projects.* Overseas Development Administration, London, 1988.

2 Initial estimates

Initial estimates should be prepared by the client or his consultant for all stages up to the invitation for tenders.

The object of any estimate, at whatever stage it is produced, is to predict the most likely cost of the project. It is essential to recognize that there is a range of possible costs within which the most likely cost lies. If the limits for the possible range of costs over the timescale of a project are plotted, an envelope similar to that shown in Fig. 2 can be produced. Fig. 2 is idealized: for example, it assumes the cost evolution of the project is continuous, but this is often not the case.

The key points are as follows.

- There is a range of possible distributions. Generally these show narrowing range and increasing certainty as the project progresses, but certainty is achieved only following settlement of the final account(s) and the auditing of all project-related costs.
- The range of possible cost is much greater during the early stages. This is because there is little information available to the estimator in terms of scope, organization and time and cost data.
- Many risks are latent in a project at its earliest stages. These often go unrecognized because the project team is intent on looking ahead rather than reviewing the present position.
- Risk decreases over the life of a project. This may not be continuous: from time to time there may be an increase in specific sources of risk or new risks may arise during the development of the project. Some of these are the result of failure to recognize their importance or presence earlier.

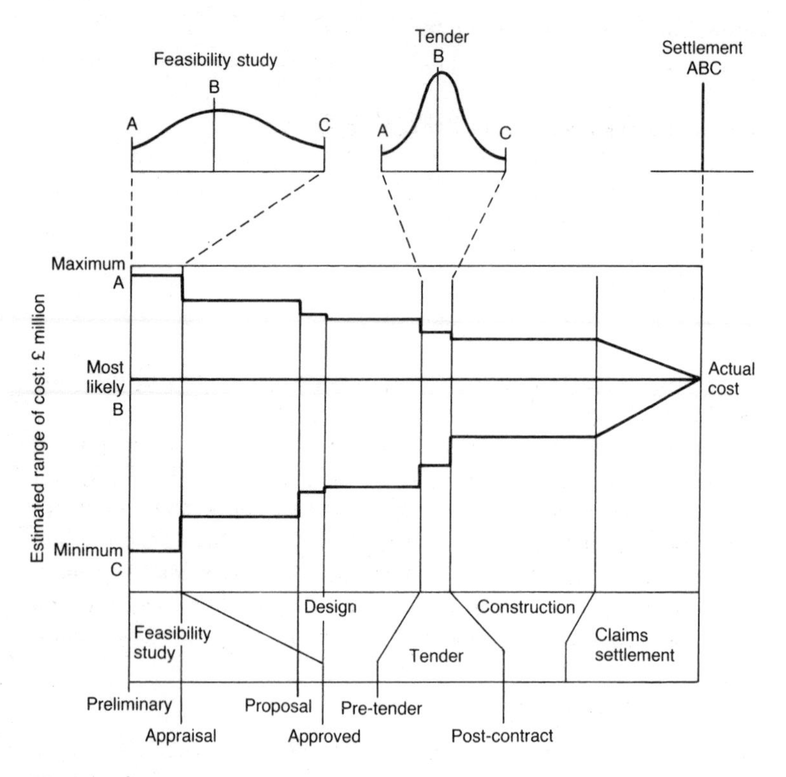

Fig. 2. Estimated envelope stepped: the estimated value does not follow a continuous evolution but is refined in stages when the estimates are produced (based on Barnes)

- The upper and lower bounds of the envelope are not absolute and in reality are not clearly defined.
- The estimated base cost plus contingencies which are likely to be spent must be close to the most likely cost from the earliest stage if the common experience of underestimation is to be avoided. It follows that, to allow for cost growth, the most likely value is closer to the maximum than to the minimum.

It can be seen that there is an immediate difficulty: there is a need to produce a realistic initial estimate but only minimal data on which to base it.

Importance of early estimates

All estimates should be prepared with care—the earliest, despite the lack of information available, no less than later detailed estimates. The first estimate that is published for review and approval has a particularly crucial role to play because it is the basis for the release of funds for further studies and estimates, and because it becomes the marker against which subsequent estimates are compared. Moreover, early estimates are important because of the need to know, for the purposes of economic appraisal, the capital cost of the project. The capital cost may not be the most sensitive variable in such an appraisal, but it can be decisive in the decision on whether to proceed with a project or to invest the funds elsewhere.

The decision by the client to sanction the project should never be based on the very first estimate. Nevertheless before significant moneys are spent on developing designs and more detailed estimates, there is a need to have a realistic indication of the likely cost of the project. By definition this must be done before a great deal of detailed information is available.

Risk in initial estimates

The earliest estimates are primarily quantifications of risk; there are few reliable data available to the estimator. To estimate effectively therefore requires the estimator not only to have access to comprehensive historical data and to be capable of choosing and applying the most appropriate technique, but also to have, in conjunction with other members of the project team, the experience to make sound judgements regarding the levels of—largely unquantifiable—risk. For many projects, this will not be an onerous task, but for projects which are in any way unusual it is essential that thorough exercises in risk identification, risk assessment and the selection of the most appropriate responses are performed.

Insofar as cost estimates are concerned, the selection of the most appropriate responses is the quantification of allowances for uncertain items, specific contingencies and general estimating tolerances. In addition the programme of the project must be reviewed carefully, and the costs of delay and the use of float or acceleration must be assessed and the appropriate contingencies included in the estimate.

This exercise is therefore time, resource and cost intensive. However, it is not required for all projects. The first step towards assessing the risk is to identify potentially high-risk projects. This can be achieved by asking simple questions, regardless of the size, complexity, novelty or value of the particular project.

- Is the project promoter's business or economy sensitive to the outcome of the project in terms of the quality of the product, capital cost and timely completion?
- Does the project require new technology or development of existing technology?
- Are there any major physical or logistical restraints such as extreme ground conditions or access problems?
- Does the project require a novel method of construction?
- Is the project large and/or extremely complex?
- Is there an extreme time constraint?
- Is the location of the project one in which the parties involved —client, consultants and contractors—are likely to be inexperienced?
- Is the project sensitive to regulatory changes?
- Is the project in a developing country?

Testing new projects against such criteria is the first step towards improving the realism of initial estimates. If the result of such a test is positive (i.e. the project is inherently risky), further and detailed assessment must be performed of the potential sources of risk and their likely impact on the project. This should be undertaken alongside the preparation of detailed cost estimates and programmes, and is crucial in supporting the estimating process by quantifying the impact of these risks in addition to the risks arising from uncertainties in scope, programme and cost/price data.

Two types of initial estimate can be identified.

- *Preliminary estimate*—the quick estimate which is needed at the earliest stage when no design is available and there is only the barest statement of the capacity/size of the project. This estimate may be used as a provisional marker in the forward capital budget, in which case it must be escalated from the base date to give a cash estimate.

- *Appraisal estimate*—estimates of alternative schemes which are under consideration in the feasibility study, carried out during the appraisal stage of the project. These estimates must be directly comparable with each other. It is therefore desirable that the same estimating technique and price base date are used. The differences between the estimates for alternative schemes will not be absolute and these estimates should not be used in the forward budget. If the estimate for the preferred scheme is to be used to refine the marker in the forward budget, then it must be escalated to a cash estimate. It must include all costs which will be charged to the project so that it is the best estimate of the total anticipated final cost. Cash estimates should also be prepared where the alternative schemes are of different durations.

Initial estimates for different contract strategies

Projects procured by the traditional civil engineering approach can be divided into a number of discrete stages. The precise number will vary from project to project, but generally five to seven are identifiable.

Projects which are procured by different contract strategies may not show these stages in the same order or may not show them all. For example, the detailed design and estimating for typical turnkey contracts is done post-contract. Fast track projects are even more compressed. To an outside observer several of the stages normally identifiable appear to have been omitted, but in reality they have been overlapped. The detailed appraisal, definition, approval and detailed (equivalent to pre-tender) estimates are performed within the implementation phase of a fast track project. Clearly this puts great pressure on the estimators to prepare a realistic estimate of the out-turn cost of the project at a very early stage. Subsequent estimates produced as the design is developed should be no greater than the initial estimate. This is the ideal for all projects, but it assumes greater importance for turnkey or fast-track projects as they are frequently let on a lump sum basis, thereby putting great emphasis on correct initial estimates and the minimizing of cost growth through the detailed design, construction and commissioning stages.

Unfortunately, it is not uncommon for technical papers and books dealing with estimating and cost management to state that an estimate prepared at a given stage of a project's development will lie within a predetermined range. This may be the case for some projects—those which may be classified as low risk—but is almost certainly not the case for projects which are in any way out of the ordinary, including many civil engineering projects.

Estimating techniques for initial estimates

Three of the estimating techniques—global, factorial and unit rates—rely on historical data of various kinds. The associated risks of using historical data in estimating are very important and the following general warning points are made.

- To obtain realistic estimates the data must be from a sufficiently large sample of similar work in a similar location constructed in similar circumstances.
- Cost data need to be related to a specific base date. For construction work carried out over a period of time, a mean date has to be chosen, e.g. two-thirds of the way through the period. For manufactured plant the easiest date to determine will probably be the delivery date, but during manufacture a mean date may be more relevant.
- Having selected the relevant base date there remains the problem of updating the cost/price data to the base date for the estimate. The only practical method of doing this is to use an inflation index, but there may not be a sufficiently specific index for the work in question. If there is not, recourse to general indices is usually made. In any event there is a limited length of time over which such updating has any credibility, particularly in times of high inflation and/or rapidly changing technology, and the estimator must review this with care.
- Overlying the general effect of inflation is the influence of the market. This will vary with the type of project being undertaken, the supplying countries and, for international projects, with the host country. The state of the market at the price date will require careful consideration before historical data can be applied credibly to a later or future date.

Global technique

The broadest category of technique is the global technique. This relies on libraries of achieved costs of similar projects related to the overall size or capacity of the asset provided. It is also known as rule of thumb or ballpark estimating. Example costs are

- cost per square metre of building floor area or per cubic metre of building volume
- cost per megawatt capacity of power stations
- cost per metre/kilometre of roads/motorways
- cost per tonne of output for process plants.

The technique relies entirely on historical data and therefore must be used in conjunction with inflation indices and a judgement of trends in price levels (i.e. the market influence) to allow for the envisaged timing of the project.

The use of this type of updated historical data for estimating is beset with danger, especially in relation to inflation, as outlined generally, but more specifically because of

- different definitions of what costs are included, e.g.
  - engineering fees and expenses of consultants/contractors/the client, including design, construction supervision, procurement and commissioning
  - final accounts of all contracts, including settlements of claims and any *ex gratia* payments
  - land
  - directly supplied plant and fittings
  - transport costs
  - financing costs
  - taxes, duties etc.

- different definitions of measurement of the unit of capacity, e.g.
  - square metre of building floor area: is the area measured inside or outside the external walls?
  - cubic metre of building volume: is the height measured from the top of the ground floor or the top of the foundations?
  - metre/kilometre of motorway: does an overall average estimate include *pro rata* costs of interchanges or should these be estimated separately?
  - associated infrastructure (e.g. transmission links/roads for power stations): is this included in the unit?

- not comparing like with like, e.g.
  - differing standards of quality, such as different construction standards, and building regulations
  - process plants on greenfield sites and on established sites

  - different terrain and ground conditions, such as roads across flat plains compared with mountainous regions
  - different logistics depending on site location
  - scope of work differences, such as power stations with or without offices, workshops and stores
  - item prices taken out of total contract prices (especially turnkey) which may be distorted by front end loading to improve the contractor's cash flow, especially for hard currency items
- inflation
  - different cost base dates (as already noted, it is essential to record the mean base date for the achieved cost and to use appropriate indices in making adjustments to the forecast date required)
- market factors
  - competition for resources during periods of high activity
  - developing technology which may influence unit costs.

Many of these items are obvious, but for projects that do not have continuous gestation and may involve several different parties from time to time, it is essential that they are thoroughly checked for. A scrutiny of all these dangers, especially the effects of inflation, must be made before any reliance can be placed on data of this type. It follows that the most reliable data banks are those maintained for a specific organization where there is confidence in the management of the data. The wider the source of the data, the greater is the risk of differences in definition.

However, so long as care is taken in the choice of data, a global technique is probably as reliable as an over-hasty estimate assembled from more detailed unit rates or other detailed prices drawn from separate, unrelated sources and applied to guestimates of quantities.

Factorial technique

The factorial technique is used widely for process plants, power stations and so on where the core of the project consists of major items of plant which can be specified relatively easily and for which current budget prices can be obtained from potential suppliers. The technique provides factors for a comprehensive list of peripheral costs such as pipework, electrics, instruments, structure and foundations. The estimate for each peripheral item will be the product of its factor and the estimate for the main plant items.

The factorial technique does not require a detailed programme; nevertheless, one should be prepared to identify problems of construction, lead times for deliveries of equipment, planning approvals and so on which can go undetected if the technique is applied in a purely arithmetical way.

An explanation of the technique for process plants is given by Barnes.* The success of the technique depends largely on

- the reliability of the budget prices for the main plant items; the estimator is still required to make a judgement on the value to include in his estimate depending on the state of the market and the firmness of the specification
- the reliability of the factors which should preferably be the result of long experience of similar projects by the estimator's organization; during periods of significant design development, certain factors can change rapidly
- the experience of the estimator in the use of the technique and his ability to make relevant judgements
- the adoption of the technique as a whole so that deficiencies in some areas will compensate for excesses in others.

The technique has the considerable advantage of being based predominantly on current costs; it thereby takes account of current market conditions and needs little reliance on inflation indices.

Man-hours technique

The man-hours technique is probably the original estimating technique. It is most suitable for design work and labour-intensive construction, including operations such as the fabrication and erection of piping, mechanical equipment, electrical installations and instrumentation, painting and insulation, where there exist reliable records of productivity of different trades (e.g. process plant construction and fabrication of offshore installations) per man-hour. It is also used for labour-intensive maintenance work. The total man-hours estimated for a given operation are then costed at the

---

* Barnes N.M.L. Financial control of construction. In *Control of engineering projects* (ed. S.H. Wearne), ch. 6. Edward Arnold, London, 1974.

current labour rates and added to the costs of materials and equipment. The advantages of working in current costs are gained.

The technique is similar to the operational technique. In practice it is often used without a detailed programme, on the assumption that the methods of construction will not vary from project to project. Experience has shown that where they do vary (e.g. in the capacity of heavy lifting equipment available in fabrication yards or on site) labour productivities, and consequently the total cost, can be affected significantly. A detailed programme should be prepared when using this technique to identify constraints peculiar to the project.

Unit rates technique

During initial estimates a detailed bill of quantities is not available. Instead quantities of the main items of work must be estimated and these are priced using updated rates which take account of the associated minor items. Taken to an extreme, the cruder unit rate estimates may be considered as global estimates.

Nevertheless so far as initial estimates are concerned, there is a real danger that the precision and detail that may be generated can give a misplaced level of confidence in the figures. It must not be assumed that previous work was similar and carried out in identical conditions or of the same duration. The duration of the project will have a significant effect on its cost. Many construction costs are time-related, as are the fees of design and supervisory staff, and all costs are affected by inflation. An outline programme should therefore be prepared to ensure that any unusual features which have an impact on the unit rates are identified.

Despite its shortcomings, unit rate estimating is probably the most frequently used technique for the initial stages of projects. It can result in realistic estimates when practised by experienced estimators with good judgement, access to a reliable, well-managed data bank of estimating data and the ability to assess the risk and uncertainty surrounding the work.

Operational technique

The operational (resource-cost) technique is the fundamental estimating technique. It is less suited to initial estimates because, compared with other techniques, its execution is relatively time-consuming and resource-intensive. However, estimating organizations geared up to this technique accumulate data in an operational form which enable them to prepare even preliminary estimates with some appreciation of the most obvious risks, uncertainties and special circumstances of the project.

The total cost of the work is compiled from consideration of the constituent operations or activities revealed by the method statement and programme and from the accumulated demand for resources. Labour, plant and materials are costed at current rates. The advantages of working in current costs are gained. It is the most reliable estimating technique for civil engineering work and it is frequently used by major contractors and an increasing number of consulting engineers.

It has the significant advantage, through the early identification of risks during the preparation of initial estimates, of highlighting the need for action to reduce their impact, thereby encouraging proactive management. In particular, the operational technique for estimating holds the best chance of identifying any significant risk of delay because it involves the preparation of a programme and method of construction, and an appreciation of productivities. Sensitivity analyses can be carried out to determine the most vulnerable operations and appropriate allowances can be included in the estimate.

Suitability of estimating techniques for project stages

In the early stages of a project few data are available, and so the techniques used must not require many data. This means global techniques and simple operational estimates are to be preferred. For any organization involved in the early identification of projects and forward budgeting, the availability of a reliable, well-managed, global cost data bank, together with the capability to perform the associated broader analyses, is essential.

During the appraisal stage, the key activity is comparison of alternative schemes. Hence techniques which facilitate reliable comparisons are required. Limited data will be available because numerous alternatives may exist only as conceptual designs. The appropriate techniques are likely to be

- global—notably the updated unit rate technique for the main work items
- factorial, if a proven system exists for the project type
- operational or man-hours when sufficient data are available.

Operational estimating is most appropriate when plant-intensive construction is envisaged because substantial capital and time-related operating costs will be associated with the major plant resources required.

Building projects or projects involving repetitive low-risk construction work may be suitable for the unit rate technique, which, in experienced hands can produce excellent results. Similarly, the factorial technique can yield high degrees of accuracy. The key to the success of these techniques is in the restriction of their use to less risky projects.

Practical aspects of initial estimating

The need to identify potentially high-risk projects and some of the factors which may make them so have been reviewed. It has also been shown that, for any project, initial estimates are inherently more risky than later estimates. The main sources of risk and how they can be addressed during the estimating process are now reviewed in greater detail.

The main sources of risk can lead the project towards success if they are identified early, assessed and managed correctly, or they can drive a project to failure if they are not identified or are identified too late, or if they are not assessed correctly or are mismanaged.

There are two components to any estimate

- the base estimate
- contingency allowances which are required to cover uncertainty in the base estimate.

It follows that the greater the uncertainty in the base estimate, the greater the contingency provision must be. However, experience has shown that the true extent of uncertainty in the base estimate is frequently unrecognized; this is the key factor that causes underestimates. The first step is to identify projects which give rise to greater than normal uncertainty by testing each new project against the criteria which define a high-risk project.

Next, a list of the main sources of risk should be prepared; there may be as few as five but there should not be more than 15. Main risk sources include

- the promotor
- the host government(s)

- funding
- the definition of the project
- concept and design
- local conditions
- permanent installed plant (mechanical and electrical)
- construction
- logistics
- estimating data (time and cost)
- inflation
- exchange rates
- *force majeur.*

These are both quantitative and qualitative.

Following this, each source of risk should be developed into comprehensive lists of more specific, manageable, and where possible quantifiable, risk factors. It is useful at this stage to classify both the potential impact (major or minor) and the probability of occurrence (high or low) as a guide to the most important factors, because the estimator's attention should be directed mainly to these high-impact and high-probability risks.

Organizations that construct or manage projects regularly will benefit from the preparation of generic lists of risks and assessment tables for each type of project they are involved with. These should be treated as an *aide-mémoire* and critically and thoroughly reviewed for each new project.

This process of risk assessment is iterative. During each stage decisions are taken to manage risks and reduce uncertainty through increasing definition of scope and refinement of the engineering design. These decisions give rise to secondary risks which must themselves be identified and quantified and/or managed out.

The process should be repeated at regular intervals to ensure that unanticipated risks either do not occur or are recognized early enough for them to be managed so that their impact is minimized. The cost impact of many of the factors can be assessed only through subjective (i.e. qualitative) judgements; others, including those that have an impact on the programme, can be assessed quantitatively by the use of probabilistic estimating and programming models.

Lessons of previous projects

Experience and research have shown that many projects which are described as overruns would be better described as underestimates. When it is possible to review the history of several comparable projects executed over a number of years, it can be seen that the estimates rise as experience is gained until they are accurate predictions of the final cost. It can also be seen that the global unit costs of the later projects are not significantly different from those of the early projects.

In other cases, the initial estimates may be realistic, but are reduced, only to rise as the project proceeds (Fig. 3). This may be caused by, for example

- attempting to meet a preordained and therefore arbitrarily set figure

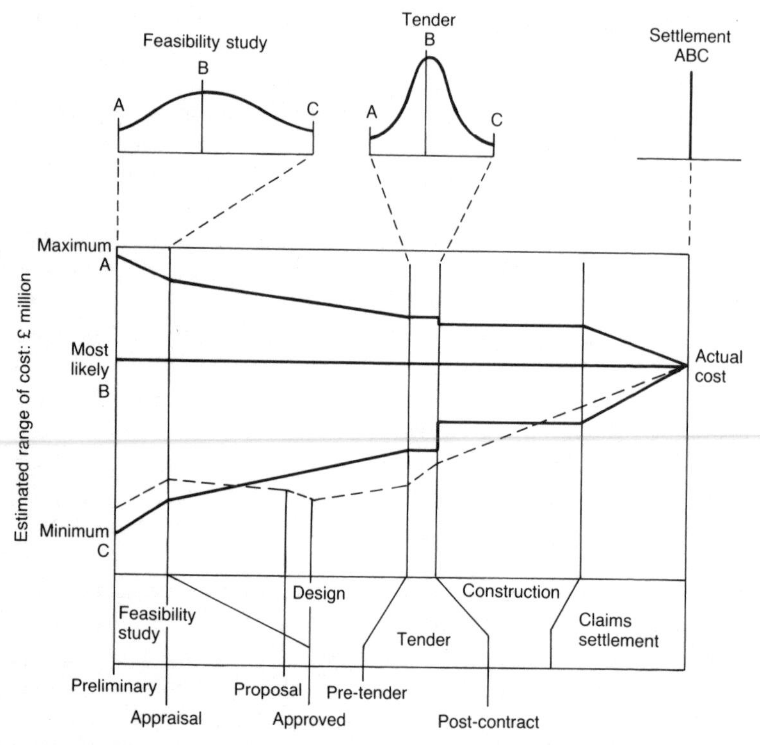

Fig. 3. Actual trend of estimates: many estimates are unrealistically low from the outset and frequently they are reduced further during the approval stage but inevitably rise during the later implementation stage (based on Barnes)

- finding that the available funding is reduced part of the way through the gestation period of the project
- listening to ill-informed outside views.

The objective of project managers and estimators must be to ensure that the initial estimates are realistic and close predictions of the final achieved cost. To this end it is essential to learn the lessons of previous projects, and incorporate them into subsequent estimates.

The main sources of risk can be identified in this way. From the point of view of cost estimating, lessons can be learned from projects which have significantly exceeded their initial estimates. Projects which fall into this category include civil engineering and building projects, petrochemical and offshore projects, power generation projects and projects in developing countries. No sector of the construction industry or size of project is immune from this problem. From the detailed reviews of the underlying causes of overspending it is possible to synthesize lists of the main categories of sources of risk, which include

- failure to plan—in the broadest sense—adequately
- failure to appreciate the organizational complexity of the project
- failure to appreciate the degree of novelty of the project to those involved
- failure to allow for external changes (e.g. to national regulations)
- underestimation of the scope of work
- underestimation of the technical difficulties
- variations and design changes
- additional costs due to delay/additional costs due to acceleration measures to meet a fixed completion date
- inflation
- unsuitable contract strategies and contracts
- reliance on tenders to define the project costs
- publication of estimates too early in the life of the project
- discontinuity in project development and lack of accountability.

Most of these items are qualitative, not quantitative, issues. Longer more detailed lists show proportionally more qualitative issues. A realistic estimate must address both the quantitative and qualitative data relevant to the costs of a project.

Inevitably, there is overlap between these main sources of risk, such as in variations and design changes caused by the need to accommodate regulatory changes. However, variations and design changes may also arise as a result of redefinition of the scope of work or clarification of technical issues. Delays may be caused by slow decision-making due to complex organization, as well as by increased work scope and variations; inflationary effects are multiplied by delay—the list goes on. However, what is clear is that the qualitative factors are not addressed directly by the most commonly used estimating techniques. These techniques tend to be mechanistic and based on quantifiable factors such as the size of the permanent works or the output of the product. Therein lies a problem: most risks are not considered rigorously or objectively. That is not to say that they are unconsidered. However, risks which should always be considered—which are largely quantitative—are frequently ill-considered. Inflation is a good example.

The lessons of previous projects reinforce the need to perform rigorous and thorough reviews of the possible impact of all sources of risk and risk factors. This takes time. The earliest estimates may have to include significant allowances to cover risk and cost growth as experienced by previous projects. Indeed for high-risk projects such allowances may dominate the estimates until late in the life of the project. The inclusion of these allowances is not due to uncertainty on the estimator's part. On the contrary, it is an empirical lesson learnt from previous projects, reflecting the risk inherent in many of them. It is tempting to many employers to seek to reduce the estimate, but if this is done arbitrarily, without reducing or managing out risks or reducing the scope, quality or functionality of the project, it is inevitable that costs will rise to realistic levels.

Review of all estimates is essential, but the reviewer should desist from any modification unless his depth of appreciation at least matches that of the original estimator. If the reason for the review is decreasing availability of funds then the whole concept of the project must be reviewed and modified. Alternatively the project must be cancelled.

The objective of minimizing cost must not be allowed to override the need for the production of realistic estimates, which are vital for effective and efficient project management and choice of investment options.

Bibliography

Thompson P.A. and Perry J.G. (eds). *Engineering construction risks —a guide to project risk analysis and risk management.* Thomas Telford, London, 1992.

Institution of Chemical Engineers. *A new guide to capital cost estimating.* IChemE, London, 1977.

Project Management Group, University of Manchester Institute of Science and Technology. *A guide to cost estimating for overseas construction projects.* Overseas Development Administration, London, 1989.

Royal Institution of Chartered Surveyors. *Cost management in engineering construction projects.* RICS, London, 1992.

Wearne S.H. *Principles of engineering management*, 2nd edn. Thomas Telford, London, 1993.

3 Estimating at the design stage

The cost of the design of a project usually has to be estimated at two stages in a project: first to enable a client to check the feasibility of his project, and second to develop more detailed estimates during the design phase to enable the designer to design the project within an agreed budget.

All design estimating or pre-tendering estimating carried out by consultants is approximate. It does not matter whether it is prepared very early on in the design process, or the pricing of a draft bill of quantities—it is still approximate and it is only the contractor's bid on which a construction contract will actually be based.

The *Oxford English Dictionary* defines the verb estimate as 'to judge tentatively or approximately the value'. The dictionary uses the word value but the consultant is not technically trying to estimate the value of some construction work. He is trying to estimate the amount of the contractor's bid for that contract and this may not be its true value, as market conditions will be brought to bear on the contractor's tender.

Forgetting commercial considerations for a moment, it is human nature to want to know what any purchase will cost before making that purchase. If you want to buy a shirt, a television or a car you can obtain the selling price in advance. If you want the shirt laundered or the television or car repaired, you can obtain the price, or a quotation, in advance of having the work carried out. Construction, however, is different. The client will know what he wants, but he needs to have the job specified and designed before he can obtain a tender, which is the quotation from the company which will actually do the work.

Clients of construction, not just because of human nature, but also for commercial reasons need to know how much a project is likely to cost long before the tender is available. They need to check economic viability, to assess financing options and obviously do not wish to waste design fees on an uneconomic project.

Feasibility estimating

The uses of estimates

Feasibility estimates have a number of uses for both clients and consultants. The main functions are as follows.

- Clients use feasibility estimates
  - to determine capital investment costs, including the cost of design, for a project
  - to determine the economic viability of a project
  - to assist with financial arrangements for a project.
- Consulting engineers use feasibility estimates
  - to aid site selection
  - to determine the design of a project
  - to assist in the choice between alternative designs
  - to predict construction costs
  - to negotiate variations
  - to assist with the adjudication of tenders.

All commercial appraisal systems require an estimate of cost in order to be able to function and all cost control systems need a yardstick against which expenditure can be measured. For construction projects this yardstick is the feasibility estimate. Everyone concerned with pre-tender estimating should therefore recognize that the estimate is the cornerstone of all cost control and financial appraisal systems.

Not only does today's client want an estimate of the cost of his project as early as possible within the design process, he also expects his consultants to design the project within the initial budget. Consultants therefore need to develop their estimate as the design proceeds, as described in chapter 2, and monitor the cost of their developing designs against the initial budget.

Estimating accuracy and influence

Many research projects have shown that the level of accuracy of feasibility estimates is not good. This should only strengthen the resolve of consultants to produce the most accurate estimates possible at the early stages of a project.

Research has also shown that the level of accuracy of the estimates prepared during the various stages of design does improve. This assists the cost control process, allowing more detailed monitoring of the capital cost as the design develops.

The decisions made in design on the size, quality and complexity of a project are the greatest influence on the final capital costs of a project. The moment a designer puts pen to paper, or instigates the first part of a CAD program, he is committing the client to some eventual cost. As a design develops more and more capital cost is committed on behalf of the client until at tender stage with the design complete, or virtually complete, the client is committed to a high percentage of the tender value. Unless a redesign is undertaken, with a consequent loss of fees and time, the consultant's ability to save money while maintaining the original design concept is very limited. Cost-saving exercises during the construction phase of a project, as well as having only a limited effect, are no substitute for cost control during the design stage.

Proposal estimating

Design cost control

The design budget estimate should confirm the appraisal estimate and set the cost limit for the capital cost of the project. As the design passes through the various stages from outline design to detailed design and working drawings so the capital cost estimate should be checked and refined to ensure that the designers are working to the budget and that the cost limit will not be exceeded or, alternatively, should be exceeded if the extra cost is shown to be a benefit.

At the same time as the consultants are controlling the capital cost of the works, they must ensure that their own design office costs are monitored and controlled. These design costs should include a small allowance for operating a capital cost control system within the office.

A consultant usually has a professional duty to take steps to try to design a project within the budget set: either a budget proposed by the client, which the consultant should have checked to ensure it is realistic, or a budget based on the consultant's own appraisal estimate. Failure to keep within a set budget can have a devastating effect on many clients.

Consultants should also endeavour to produce economical design solutions, to ensure that the client is not paying for over-design or expensive design alternatives. Taking measures to control the capital cost of a project during the design phase therefore is to keep expenditure within the amount allowed and to give the client good value for money.

Design principles that affect cost

There are many good principles of design that help produce economical designs and assist in keeping the capital cost of a project within budget. Most senior consultants are well aware of these principles and in fact take them for granted. The main points are as follows.

- *Structural shape.* The construction of curved masonry or concrete walls is relatively expensive. This is therefore not generally the most economical shape for a structure in spite of the fact that it gives the most floor area for the least amount of wall. The closer a design keeps to a square the better will be the floor to wall ratio. This is a simple concept to be kept in mind when considering all the more technical matters of any design.
- *Buildability.* Some designs are more difficult, and therefore more expensive, to construct than alternative designs which provide a very similar end result. Design draftsmen often appear unaware of the construction problems created by even the simplest of designs. Reducing the size of in situ concrete columns as they rise through a structure saves concrete and may even be aesthetically pleasing on external elevation. However, it also involves the contractor in costly reconstruction of the formwork for each lift and puts the economics of such design into question.

  A short survey of contractors' site staff by researchers at Loughborough University identified 153 examples of poor design from the point of view of buildability. Many of the examples quoted concerned reinforced in situ concrete design. A number of contractors stated that it is more economic to construct retaining walls with stepped rear faces than those with sloping rear faces. Nearly all the contractors gave examples of column beam and slab junctions where the mass of reinforcement made it virtually impossible to place the concrete. Every contractor interviewed gave an example of overcomplicated in situ concrete members that increased the cost of shuttering out of all proportion to the cost of the concrete and reinforcement in the basic member.

- *Knowing what plant is available.* Some designs require particular items of construction plant for their execution: specific items of plant possibly have to be hired for just one operation on site. The conclusion must be that designers should remember to consider the implications of their designs on construction processes.
- *Specification of materials.* Too many specification clauses ask for particular materials by trade name rather than specifying that materials should comply with British Standards and allowing the contractor to meet the specification with the most economic suitable material. Where a material is not a visible component of the finished structure there is often no good reason for specifying a specific named material and thus possibly denying the client the benefit of an economic alternative.

By practising these various basic principles of good design a consultant engineer lays the foundation of an economic design and capital cost control of a project.

Design office capital cost control

To be able to exercise cost control during the design phase of a project a design office must introduce a cost control system to be used throughout the office on all its projects. The object will be to monitor the cost of the designs produced against the budgets. The system will also assist in ensuring that the client obtains value for money and that there is a correct balance of expenditure between the various sections of the project.

- *Initial reference system.* The reference point for any cost control system is the agreed budget with the client which is often produced as the appraisal estimate. This will be the cost limit and the envisaged tender cost. The budget itself has to be broken down between the various sections of the project to provide cost targets for each section. The cost targets should be arrived at using the most appropriate estimating technique for the section of work depending on the amount of information available at this early stage of the project.

- *Checking the cost of designs*. As the design of a project progresses and the basic designs are produced for sections of the work, an estimate of the cost of the section is produced based on the actual drawings for the work involved. Estimates at this stage are usually carried out using approximate quantities, but other estimating techniques may be used if appropriate. Where the cost estimate for a section of work is within the cost target for that section, no further action is required and the design can remain as part of the contract document. However, if a cost estimate exceeds the cost target for the section, further action is required and the third stage of the system must be used.
- *Means of remedial action*. Where the cost of a section, as designed, exceeds the cost target some action must be taken to keep the overall cost within the project budget. The common forms of remedial action are
  - to redistribute the overall cost among the sections
  - to respecify items of work
  - to redesign the section
  - to use a design contingency
  - to ask the client to increase the budget.

Proposal evaluation

Once the design has been finalized the consulting engineer's estimate can be prepared, based on the level of design completed. Many consultants may also price the draft bill of quantities as part of the estimating process. By comparing the major rates with a contractor's tender any discrepancies at this level can be identified and appropriate action taken to deal with any errors.

Bibliography

Ashworth A. and Skitmore R. M. *Accuracy in estimating*. Chartered Institute of Building, London, 1983, occasional paper 27.

Corrie R. K. *Project evaluation*. Thomas Telford, London, 1990.

Royal Institution of Chartered Surveyors. *Cost management in engineering construction projects—guidance notes*. RICS, London, 1992.

4 Estimating for construction

This chapter deals with estimating at the implementation stage of a project and the preparation of pre-tender and post-contract estimates.

Competitive environment

The public sector clients—namely local and central government—with a need for public accountability have developed many of the practices for competitive tendering and private clients have tended to follow suit. Thus most construction contracts are awarded after several contractors have prepared estimates and submitted tenders. A tender price, or bid price, can be described as

bid price = base cost + mark-up

This conveniently divides the bid into two elements. The base cost should be assessed by the careful assembly of cost data, the careful selection of method and the resources required, the determination of the likely output or productivity factors and the synthesizing of calculations to produce the direct cost estimate. The mark-up is an allowance for overheads, profit and risk. The determination of the mark-up allowance requires commercial judgement of the current market. To permit these commercial judgements to be made soundly, the base cost must be estimated accurately. This requirement highlights the importance of the base cost estimate. The importance of this estimate is further emphasized by the proportion of the total bid that it represents: it is the largest element. Thus any errors or variability in the estimate could well be significantly greater than the adjustments made in the mark-up. For these reasons all contractors have well-established systems, both manual and computerized, for producing reliable, consistent and accurate estimates.

Importance of estimating

One of the keys to the commercial success of a construction company is its ability to estimate the costs of construction work and to construct that work for the estimated cost or less. In competitive tendering the tender price is based on the estimated cost which may represent as much as 90% of the final total. The additions and allowances added for risks, overheads and profits are the much smaller proportion of the tender price and therefore are less significant in the determination of which contractor will be the lowest bidder. The relative importance of the elements, based on typical averages in the UK for 1970–90, in a bid are

• base cost		
	labour plant materials subcontractors	90%
	site on-costs	3%
• mark-up		
	company overheads	4%
	profit	3%
	risk	0%
	bid	100%

To produce a satisfactory contract the contractor must have a bid that is low enough to win the contract and yet high enough to provide the opportunity for making a profit. Given the low profit levels experienced in UK construction this is an extremely narrow band within which to work. An error of 1–2% in estimating represents a 30–60% loss of profit. It leaves little room to cope with the vagaries of site management. In arriving at the value of each of the elements in the estimate, the contractor and his estimating team combine known and reasonably reliable data with subjective judgements. The following outlines the elements where judgement is applied.

- Base cost estimate
  - *Labour*: established from the assembled costs and recorded productivity outputs; the areas of judgement are mainly in the adjustments to productivity factors.
  - *Plant*: established from quotations from plant hire

companies or the calculated costs of operating plant together with recorded productivity factors; the areas of judgement are mainly in the adjustments to productivity factors and in the assessment of idle time.
  - *Materials*: established from quotations from suppliers together with additions for storage, handling and wastage; the areas of judgement are mainly in these additions.
  - *Subcontracts*: established from quotations from subcontractors with additions for the services the main contractors would provide; these services can normally be assessed accurately; the main risk in using subcontractors is potential disruption to the construction work due to the problems in co-ordination.
  - *On-costs*: established from the assessment of the site staff and other general support facilities required; this can normally be done with accuracy.

- Mark-up
  - *Overheads*: established from a calculation of the head office budget and the company's forecast project turnover.
  - *Profit*: established from a calculation of the minimum profit a company needs to service its debts, satisfy its shareholders, re-invest in the company and pay its taxes; however, the main determinant of the profit element is a judgement on what the market will bear.
  - *Risk*: established from an assessment of the risks involved in the project and thus largely a subjective element.

Thus the structure of a bid could be represented as

- *base cost*: calculated largely with inputs that involve some judgement

- *mark-up*: predominantly by judgemental elements.

The assessment of these data has led contractors to a clear set of conclusions.

- Regardless of what the market conditions are, the calculated elements are by far the largest. These calculated elements contain input data that contain subjective assessments.
- Within these calculated elements the largest is the direct costs of labour, plant, materials and subcontractors. Therefore every

step must be taken to ensure that these are calculated with the greatest accuracy in order to minimize the variability in the direct cost estimate.

- The profit and risk element, although important in the bid equation, is much smaller than the direct cost element; variations in the direct cost estimate can be much larger than any adjustment considered in the profit and risk element.

Tendering process

The process undertaken to produce a competitive bid for a project is shown in Fig. 4. It consists of the following stages

- the decision to tender
- programming the estimate
- a preliminary project study
- a project study including a production of method statement
- preparing the base cost estimate
    - collecting or calculating cost information
    - calculating direct costs
    - calculating site overheads or on-costs
- preparing reports for tender adjudication
- the adjudication process.

The first five of the stages are now considered; the last two are described in chapter 6.

In all sizeable companies the tendering process is developed and refined until it is efficient in terms of the number of consistent and reliable estimates produced per estimator. Efficiency is required in both depressed and buoyant markets. In depressed markets the success rate declines and more tenders, and hence estimates, need to be produced to maintain a company's turnover. In buoyant markets the increased bidding opportunities need evaluation to ensure that a company makes the best use of its opportunities.

Decision to tender

All companies should have a corporate plan. Some companies' plans are more detailed than others. The corporate plan will give details of a company's turnover targets broken down into various divisions or sectors of work. Against this corporate plan senior managers will take the decision to bid for a specific contract. The factors considered will include

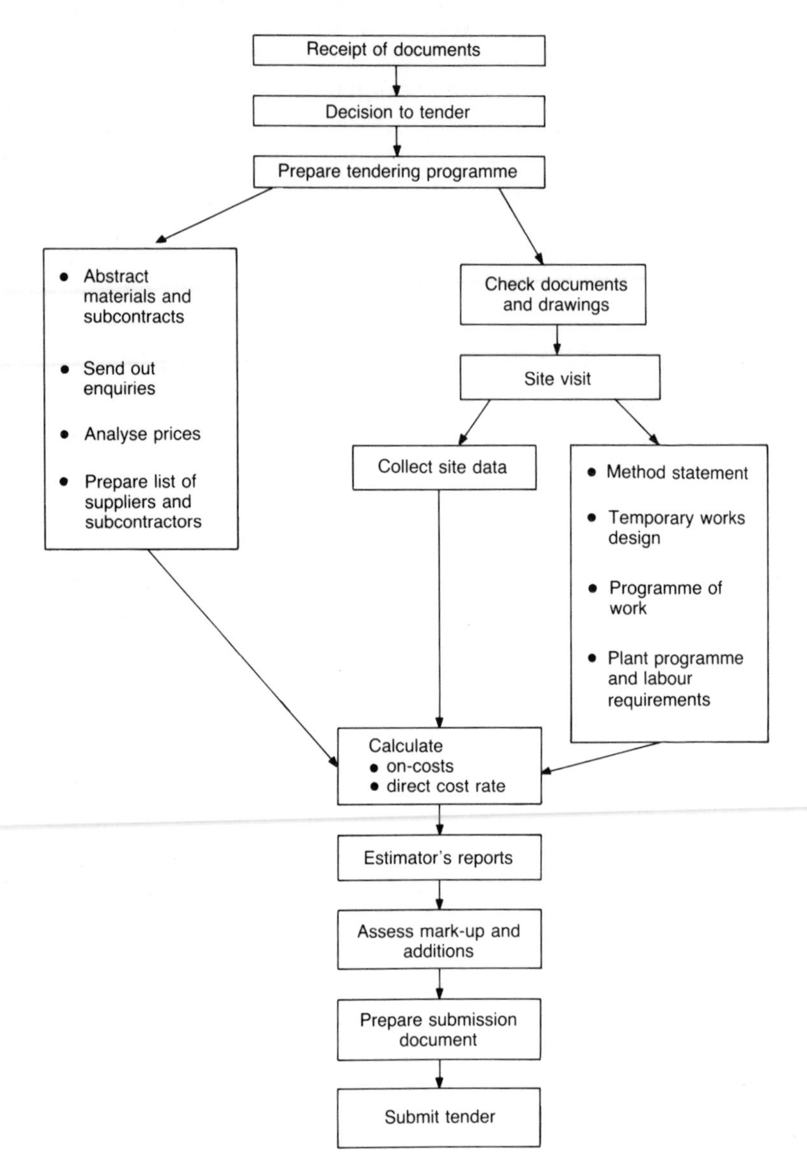

Fig. 4. Preparation of tender source

- the potential contribution of the contract to the company's turnover in a particular sector, the overhead recovery and the anticipated profit
- the likely demands of the contract on the company's financial resources
- the company's resources available
- the type of work
- the location
- the client
- the contract details.

Companies will avoid contracts that are too large for their size, beyond their experience range, those that are likely to stretch their available resources, including cash, or are well outside their normal geographical area of operation, and contracts that have unusually onerous conditions of contract. In taking the decision to tender a company's senior managers are making judgements that balance market opportunities and risk.

The actual decision to tender may be taken at three stages within the tendering process. The first is during the pre-selection stage, if this is used. Pre-selection allows contractors to examine brief details of the project in order to allow them to indicate their willingness to produce a complete tender. This process is used on large projects where producing a tender will involve considerable time and cost to the contractor.

Pre-selection information provided to support the decision to tender includes

- the names of the Engineer with supervisory duties (if any)
- the names of the client, the prospective Engineer, Architect or equivalent
- the location of the site
- a general description of the work involved
- the approximate cost range of the project
- details of any nominated subcontractors for major items
- the form of contract to be used
- the procedure to be adopted in examining and correcting priced bills
- whether the contract is to be under seal or under hand
- the anticipated date for possession of the site
- the period for completion of the works

- the approximate date for the dispatch of tender documents
- the duration of the tender period
- the period for which the tender is to remain open
- the anticipated cost of liquidated damages
- details of any bond requirements
- any particular conditions relating to the contract.

Part of the contractor's consideration, particularly with respect to a private client, will be determination of the financial strength of the client. The history of the industry is littered with tales of private clients who have failed to pay and have offered contractors numerous alternatives such as shares in hotels and squash clubs, and even part-ownership of a racehorse! Recurring doubts as to the financial security of clients has encouraged contractors to develop skills in financial engineering. This benefits clients for whom contractors can often make financial arrangements on terms more favourable than the clients can secure, and at the same time it ensures payment for the contractor. This type of action can be seen as both providing an attractive service to a client, thereby increasing a company's competitive advantage, and reducing risk.

If a contractor indicates his willingness to submit a tender at the pre-selection stage and he is invited to do so by the client's representatives he will normally proceed through the stages outlined. However, he has two further opportunities when he can withdraw from the tendering process

- after a careful examination of the full contract documentation
- after the estimate has been prepared and the tender is ready for submission.

Planning the production of the estimate

Having taken the decision to tender for a project and having received the complete contract documentation the estimator's first task is to establish a schedule of estimating activities against key dates so that the process can be monitored and controlled effectively. This is vital as the submission date and time are precisely defined and the time available to produce the tender is limited.

This internal management action ensures that no in-company errors put at risk the estimating and tendering procedure and to ensure that each stage in the process is given adequate and due consideration. By these actions the company minimizes the risks of leaving decisions too late and being forced to substitute guestimates for estimates.

For small and medium-sized projects the planning of the estimate will be in the form of a schedule of activities with dates when key decisions must be made. On larger projects, in order to monitor and control bid production the estimating director will demand that a full bar chart programme covering the tender period is produced and used.

Project study

The study of the project may be divided into two stages.

Stage I

The purpose of the first stage of a project study is to aggregate the material, plant and potential subcontractor items within the project, so that enquiries for quotations can be made as early in the process as possible. The first stage project study ascertains

- the principal quantities of the work
- an approximate estimate
- the items to be subcontracted
- the materials and plant for which quotes are required
- critical dates for actions by subcontractors and suppliers
- whether or not design alternatives should be considered.

The four main cost headings of any estimate are labour, plant, materials and subcontractors. Quotations are required for materials and subcontractors. Typically, materials account for 30–60% of a project's value; subcontractors can typically account for 20–40%. If in an estimate the materials account for 50% of the project cost and the materials element is entered inaccurately because of a lack of quotations, the consequences are not difficult to calculate.

With respect to materials, the main difference between companies bidding for the same work is their skill in buying, not in their efficiency in constructing. It is therefore critical to make enquiries of suppliers early to ensure that most prices are obtained in time for inclusion in the estimate or—ideally—in time for negotiation to take place with a view to obtaining the best possible price. This early issue of enquiries is important both for competitive reasons and the reduction of risk.

Stage II

The second stage of the project study incorporates a complete review of the construction work which results in the method statement and pre-tender programme. Because of the interrelation of planning and estimating at this stage, both functions are involved.

The study includes a detailed review of the contract documentation, a site visit and the preparation of a method statement determining how the project will be constructed.

Contract documents. Although the preliminary study will have identified any aspects of the project which require clarification and the approximate principal quantities involved, the appraisal now undertaken refines this initial information for detailed analysis. It also requires a detailed listing of any unusual conditions contained in the contract.

Site visit. A site visit is normally arranged through the client's representatives. In building contracts such visits are not always arranged, in which case the contractor will make his own arrangements for a visit. The purpose of the site visit is to prepare a report containing

- a description of the site
- the position of existing services
- a description of ground conditions
- details of available access and restrictions
- details of any potential security problems
- topographical details
- details of any temporary work or demolition to adjoining buildings
- details of facilities for the disposal of spoil
- an assessment of the availability of labour
- any other intelligence that can be acquired with regard to the contract or competitors.

This report is used in the production of the method statement. It is also essential in that the major forms of contract require the contractor to take account of circumstances that could be foreseen by an experienced contractor.

Preparation of method statement. In large companies the method statement is usually prepared jointly by the estimator and the planner. This statement specifies the methods and procedures to be adopted in the execution of the work and defines the initial estimates of activity durations, sequence and interactions in the pre-tender construction programme.

The preparation of the method statement involves liaison and consultation with site staff, plant managers and temporary works designers. It is at this stage that design alternatives to the works may be considered. This process produces a description of the proposed construction method based on the pre-tender programme and the overall quantities of labour, plant and temporary materials required. These may be reworked several times before the tender is finally submitted.

Preparing the base cost estimate

The estimator's main task is to determine the most likely cost to the contractor in undertaking the work described in the contract documents. The preparation of the base cost estimate comprises collecting or calculating cost information, calculating the base cost and calculating the site overheads or on-costs.

The mark-up, which includes allowances for head office overheads, profit and risk, is considered separately by senior managers and added to the estimated direct cost to give the tender sum. The assessment of mark up is described later.

Collection of cost information

The cost information required by the estimator comprises details of labour, plant, materials and subcontractors.

Labour. The cost of the company's own labour is usually calculated either per hour, per shift or per week. The cost to the company of employing their own labour is greater than that paid directly to the employee. The elements in the calculation of labour costs include

- basic rate
- overtime payments
- rates for additional duties
- bonus
- tool money
- travel moneys
- holiday stamps and death benefit
- sick pay
- National Insurance

- training levies
- employer's and public liability insurance
- allowances for severance pay
- allowances for supervision.

The basic data for the calculation are contained in the working rule agreements. However, estimators have to make a number of assumptions that cannot always be supported by calculations. These include the bonus, whether guaranteed or based on productivity, which may be the most significant element and may not be precisely determined until on-site negotiations have taken place. Other assumptions include the hours lost due to inclement weather and time off due to sickness.

Plant. Plant may be obtained for a contract either internally or externally. Quotations for the plant required are therefore obtained from external hirers or from the company's own plant department.

It is rare for UK contractors in the domestic market to calculate the all-in plant rate from first principles. This calculation is usually undertaken by the plant hire company. Such calculations are undertaken for overseas contracts.

Further information may be obtained from the bibliography.

Materials. Materials form a significant percentage—typically 30–60%—of most projects. Up to date quotes are usually obtained for most material items. The material enquiries cover information on the required quantities, the delivery rates, the specification and any terms and conditions on which the quotation is being sought.

The materials quotation received would only normally include delivery to site; the estimator would add allowances for unloading, storage, handling and wastage. These allowances, if inaccurately judged, could lead to significant underestimating of material costs.

One aspect that estimators must take care to check is that the supplier has quoted for the materials specified and not for available alternatives.

Subcontractors. In the first stage of the project study the estimator will have identified items of the work to be subcontracted. The decision to subcontract depends on the specialization of the work involved. In undertaking large projects or projects of a unique nature, a company may decide to subcontract work normally undertaken by themselves in order to gain access to additional resources and spread the financial risk.

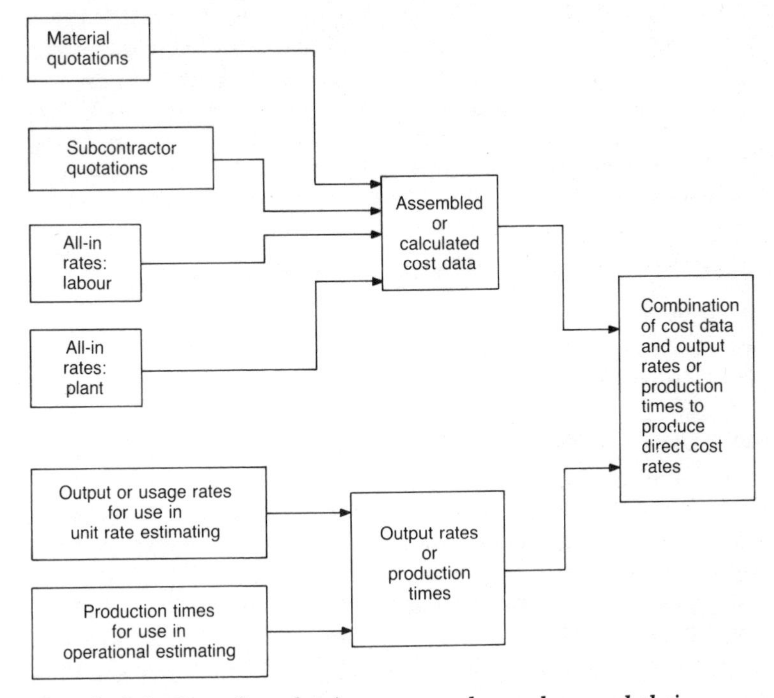

Fig. 5. Selection of production rates and cost data and their combination to produce cost rates

The quotation of each subcontractor has to be adjusted for attendance and other allowances. Each subcontract quotation will vary by the amount required for attendance (e.g. access scaffolding, crane and air compressors) which is expected to be needed. The estimator must therefore allow for this in the final subcontract rate included in the estimate.

The choice of subcontractor is important. Most companies maintain a list of approved subcontractors for various trades.

Calculating base cost rates

A direct cost rate is the cost to the contractor of undertaking the work described. The cost will include the costs of all labour, plant, material and subcontractors employed in constructing the section of work concerned. The real art of estimating is in the selection of the detailed resources of labour, plant, materials and subcontractors. An estimator's skill is in being able to define the configuration of labour, plant and materials required to undertake specific items of work. Having selected the resources, the amount of these resources required to undertake the work must be estimated —in hours for labour and plant and in quantities for materials. Having determined the resources and usage (time or quantity), the remaining task is to combine this information with the collected or calculated costs to produce a direct cost rate as shown in Fig. 5.

The main calculation techniques adopted by the estimator include

- unit rate estimating
- operational estimating
- spot rates.

Unit rate estimating

Unit rate estimating is the selection of the resources (labour, plant and materials) required and the selection of output or usage rates for these resources. Thus for each resource the calculation is the output or usage rate combined with unit cost. An output rate is work quantity per hour (e.g. labour for steel-fixing tonnes/hour) and a usage rate is time or resource quantity to do a fixed quantity of work (e.g. hours of labour to fix/tonne of steel). This method of calculating gives rates that may be directly entered against bill item rates and is therefore commonly used.

Example of unit rate estimating calculation

The rate for the provision and fixing of 16 mm dia. reinforcement includes

- material purchase costs
- transportation costs
- bending and fixing costs
- material wastage allowances
- allowances for additional materials (spacers and tie wire).

For 16 mm dia. reinforcement

Material purchase cost	=	£320/tonne
Wastage allowance	=	5% (1·05)
Tie wire and spacer cost	=	2% material cost (1·02)
Cut and bending rate	=	15 hours/tonne

Fixing rate	=	20 hours/tonne
Bar bending machinery	=	included within site on-costs
Site transportation	=	included within site on-costs
Labour rate for steel fixer	=	£6.50/hour
Total material cost	=	(1·0 × £320.00 × 1·05) × 1·02
	=	£342.72/tonne
Total labour cost	=	(15 + 20) × £6.50
	=	£227.5/tonne
Total rate for the provision and fixing of 16 mm reinforcement	=	£342.72 + £227.5
	=	£570.22/tonne

Operational estimating

Operational estimating is based on the calculation of the total quantity of work involved in an operation (e.g. the excavation) and the estimate of the total elapsed time of the operation (e.g. weeks) and a combination of these and the selected resources (e.g. the labour and plant in the excavation team). This method of calculation is favoured by estimators involved in plant-dominated work such as excavation and concrete placing. It is considered that this method of estimating, which links well with planning, deals more accurately with plant output which is subject to non-productive travel time, idle time and down time.

Example of operational estimating calculation

A total of 3000 000 m^3 of material is to be dredged from a harbour basin and disposed of on-land adjacent to the coastline. From the programme of works, dredging, reclamation and slope protection extends over a six-month period.

Dredging of material

Plant costs

Dredger at £100 000/month for 6 months = £600 000
Fuel cost is estimated at 5% of the dredger costs = £30 000

Labour costs

Six crew will be required throughout the period. Payment for this labour will be on a contract basis at a total cost of £12 500 per man including all shift working allowances and subsistence costs.

Labour cost = £75 000

Total dredging costs = £600 000 + £30 000 + £75 000 = £705 000

Disposal of material
Plant costs
D6 bulldozer at £7500/month for 6 months = £45 000
Fuel cost is estimated at 5% of the bulldozer cost = £11 250

Labour costs
Operator costs for the bulldozer will be at the rate of £6.50 per hour for an average 70 hours per week. The cost of a banksman will be £5.50 per hour for an average 70 hour week.

Labour cost = (£6.50 + £5.50) × 70 × 26 = £21 840

Total cost of disposal = £45 000 + £11 250 + £21 840 = £78 090

Cost of dredging and disposal = £705 000 + £78 090 = £783 090

Cost of dredging and disposal per m^3 of material = £783 090/£3000 000 = £0.261.

Spot rates

Spot rates, or gash rates, are direct cost estimates not based on calculation but entered into the estimate based solely on the estimator's judgement. Examples and more detailed descriptions of these calculations are given in the bibliography.

In any contract a number of items (e.g. the excavation and concrete placing) are likely to carry the largest value. These will be estimated by what is considered to be the more accurate method of operational estimating. Other items will be estimated by unit rates and the items whose contribution to the value of the project are least significant will be processed by producing spot rates. In this way the most cost-significant items (i.e. those carrying the largest value) will be estimated most accurately and given the greatest attention. It is the results of the estimator's calculation of these items that will determine whether or not the contractor wins the contract and, if he does, whether or not the contractor can make the planned profit.

Calculation of site overheads or on-costs

The estimator also calculates the site overheads, or on-costs, for the project. These include allowances for

- site management and supervision
- clerical staff and general employees
- plant
- transport
- scaffolding
- miscellaneous labour
- accommodation
- temporary works and services
- general items
- commissioning and hand-over
- sundry requirements.

These items are usually given on a check-list prepared by the company which gives guidelines on the appropriate allowances to be made in each category. The estimator determines the total costs.

The calculated site overheads are often allocated to the preliminary section of the bill of quantities but may be allocated to the bill item rates according to the contract programme or cash flow requirements. The following example shows the calculation of clerical staff and general employees for a project.

Example of calculation of site overhead costs

The site will require the following clerical staff and other general employees over the 18 months' duration of the project.

Office manager	1 at a salary of £16 500 per annum
Cashier	1 at an hourly rate of £6.50 per hour
Wages clerk	1 at an hourly rate of £5.50 per hour
Storeman	1 at an hourly rate of £6.50 per hour
Plant checker	1 at an hourly rate of £5.50 per hour
Typist/telephonist	2 at an hourly rate of £5.00 per hour
Safety officer	1 at a salary of £18 500 per annum (this cost shared with three other projects)
Security staff	The provision of security staff is obtained from an outside company at a total cost of £3475 per month.

Assume

- all hourly rates quoted are all-in rates
- all hourly paid employees will work an average of 65 hours per week with the exception of the typist and telephonist who will work 40 hours
- the total cost of salaried staff per annum = 2·10 × annual salary
- all recruitment costs and so on are included in the above rates.

The total cost of the clerical staff and general employees over the 18 months' duration of the project will be as follows.

Salaried staff		
1 × 2·10 × £16 500 × 1·5	=	£51 975
1 × 2·10 × £18 500 × 1·5 × 0·25	=	£14 568
Hourly-paid staff		
2 × £6.50 × 65 × 78	=	£65 910
2 × £5.50 × 65 × 78	=	£55 770
2 × £5.00 × 40 × 78	=	£31 200
Security staff		
Subcontracted cost £3475 per month × 18	=	£62 550
Total cost	=	£281 973

Bibliography

Civil Engineering Construction Conciliation Board for Great Britain. *Constitution and Working Rule Agreement*. Federation of Civil Engineering Contractors, London, 1992.

Harris F. and McCaffer R. *Modern construction management*, 3rd edn. BSP, Oxford, 1989.

McCaffer R. and Baldwin A. N. *Estimating and tendering for civil engineering works*, 2nd edn. BSP, Oxford, 1991.

5 Detail estimating of process plant costs

The fundamental difference between detail estimating for process plant and that for building and civil engineering works is that the base cost of the process estimate has to be obtained from material and equipment suppliers, plant vendors and subcontractors. The aim of the estimator should always to be to obtain firm quotations together with guaranteed delivery dates and installation schedules from suppliers and subcontractors because these form most (commonly 80%) of the base cost of a project.

Global estimates can be made by comparative means for similar type plants based on throughput, for example, as described in chapter 2; detailed and competitive estimates are costly and time-consuming affairs because

- materials and specialist equipment are bought on a worldwide basis and prices are subject to demand and currency fluctuations
- the construction of process plants is labour-intensive and construction costs are reflected by layout and location.

It is therefore essential to have a good definition of the scheme (i.e. layout, piping and instrument diagrams, equipment lists and material specifications) before firm prices are sought.

The main process plant contractor—the engineering contractor —usually carries out the design, procurement and management functions. This makes up most of the remaining cost (about 20%). Construction is undertaken by specialist subcontractors. The engineering contractor's first task is to quantify his own base costs as accurately as possible. This can be based on a comprehensive man-hour estimate for detailed design, procurement, project management and site supervision.

Estimating function

One common form of contract for process plants is the fixed price turnkey contract in which the engineering contractor agrees to deliver the plant ready for operation by a fixed date. This type of contract is best applied where the scope of services to be provided is well defined. It is particularly suitable for standard-type plants where the influence of the client during the design and manufacture should be minimal.

To support a fixed price proposal the contractor must strive to obtain an estimate of his costs to within an accuracy of ±10%, i.e. it is essential to have a good fix on the total base costs before considering the commercial bid.

Objectives of the parties

The client's prime objectives when choosing the fixed price contract option are

- confidence—to gain evidence that the bidder can perform
- cost—to identify and fix total costs
- quality—to maximize quality at the lowest price
- risk—to minimize risk, generally by offsetting to the contractor
- deliverables—to define deliverables
- programme—to define and fix a realistic project programme
- design—to retain some ability to influence the detail design (although this conflicts with minimum variation).

A fundamental question to be addressed is whether or not these objectives are practicable and achievable, given the scope, definition and development of the project at the time of the enquiry. Too often this question appears to be ignored.

The contractor's prime objectives are

- risk—to minimize risks by defining scope and deliverables, and to offset cost risk on to suppliers if possible
- cost—to quote the lowest price
- quality—to maximize quality at the lowest price
- programme—to offer a practical deliverable programme
- standards—to control standards and minimize special requirements
- engineering—to get it right first time, and thereby minimize variations

- design—to retain control and minimize client interference
- profit—to complete the contract at the forecast margin or better.

The question that the contractor must address is whether or not he can achieve these aims with the information available to him and within the bid period.

Technical estimates

It is important that the role and responsibilities of the estimator on a fixed price contract are clearly understood. Using the information contained within the enquiry document, and additional data generated within the engineering disciplines, the estimator will prepare a basic cost estimate of the plant. It must be appreciated that a considerable amount of engineering work must be completed if a ±10% estimate has to be prepared.

Initially the base, or net, cost will be established for

- project and engineering services—prepared in-house
- plant—quotes from vendors
- equipment—quotes from vendors
- bulk materials, e.g. cabling—quotes from vendors
- fabrication—quotes from specialist contractors
- installation—quotes from specialist contractors
- construction costs—quotes from specialist contractors.

To fulfil all the criteria, the cost of preparing a fixed price estimate can be as high as 1–3% of the total cost and take up to six months to prepare. Typically three months are allowed by most clients. This is really too short a period to enable all the requirements for the completion of a fixed price estimate to be fulfilled.

The client should be aware of the work undertaken by a contractor in tendering. The bid must be received, acknowledged and assessed. On the decision to proceed, the information must be copied and circulated to all relevant departments. Before work can be started, the bidding strategy needs to be prepared, the estimating strategy and programme have to be determined, the start-up meeting has to be held and all disciplines have to be informed of their input. However, in the real world the ideal is rarely achieved, and enquiry packages are no exception. There are several problems in fixed price estimating with respect to process definition and engineering definition.

On package plant the contractor may have responsibility for preparing process definition and design. The control philosophy must be clear in order to minimize change at the detailed engineering phase. This is not always the case. If fundamental areas such as process design and control are left open to interpretation at the bid stage, the result may well be a different plant being costed by each bidder. However, it is the client's responsibility to ensure that clear definitions are given in the enquiry documents.

The key stages in the process definition are

- clear definition of the scope and the deliverables
- definition of the input and output requirements, volume, pressures and temperature
- accurate sizing of all major equipment
- preparation of process and mechanical data sheets
- inquiries for major equipment needed
- inquiries for secondary materials and bulk quantities
- definition of the scope of work for major subcontracts and budget prices
- preparation of the method statement.

The client is usually responsible for the first two of these items by defining the basic process flow diagram and outline plant layout, although it is not uncommon for the contractor to be retained to assist at this stage. This procedure is known as a front end study and forms the basis of the bid package for competitive tendering. It is typically worth 1–2% of the total contract price.

In practice problems are overcome, not by directing questions to the client because he is unlikely to have the answers at that time, but by making decisions, assumptions and qualifications relating to the conditions and the client's requirements, and then proceeding on that basis. The danger is that everything from this point in the estimating procedure onwards is derived from the process definition, which could be ill-defined.

In general the project scope is usually clear with respect to the engineering, project management and construction services required. Areas which are not always well defined but are related to the physical scope of the work include

- boundaries
- site preparation

- infrastructure
- provision of utilities
- interface with existing plant
- hot/cold tie-ins
- demolition and removal of existing facility
- condition of reuseable plant from an existing installation.

Some questions must be dealt with at the early stages of a project. A site visit can be used to clarify many points and some questions can be addressed directly to the client. However, in a competitive situation it may not always be in the interest of the contractor to raise questions which relate to potential cost savings as the client's practice of circulating answers to all bidders will inform his competitors. In practice, this situation may be overcome by assumptions, by interpretation of the client's requirements and by qualifications relating to included scope of supply. None of these solutions is in the long-term interest of the client.

Commercial estimates

The estimator then prepares the commercial estimate. Profit and commercial contingency will be added under the general guidance of the organization's business management. Cash flow analyses will be prepared, potential cash lock-ups considered and any other financing costs added. The commercial estimate is then presented for review, amendment and final approval.

Commercial conditions must be acceptable to both parties if a contract is to be agreed. When liability clauses are tabled, as happens frequently, this can expose the contractor to risks in excess of his own potential contract value; this demonstrates the incompatibility of the risk/reward relationship.

Key factors which should be defined in the commercial estimate include

- payment terms
- insurances
- retentions
- performance guarantees (these are particularly important if the contractor owns the process licence)
- taxes
- warranties
- liquidated damages

- maximum liabilities
- detailed contract terms
- downstream liabilities (which have an adverse impact on any other plant on the same site)
- programme.

A key issue which will affect the price is the contract programme, i.e. time is money. This will influence the contractor's man-hours, inflation costs and the imposition of liquidated damages.

When preparing his contract programme it is important for the contractor to take due account of the delivery times for specialist equipment and to note the site working hours on which the installation subcontractors have bid.

In the case where the client will not accept a realistic (achievable) programme the contractor has to make his own commercial decision on whether or not to accept the client's tighter completion schedule by weighing the risks against the rewards offered under the contract.

The absence of commercial terms or the inclusion of unreasonable conditions will result in each subcontractor tabling counter-proposals. Each one will be different, leaving the client with the difficult, expensive and time-consuming task of reconciling the bids.

Allowances and contingency

By definition the fixed price bid should cover the full cost of plant, if no variations are generated. Therefore, each cost element needs to be examined and, if necessary, adjusted to achieve this goal. This adjustment consists of two stages: the application of allowances and the consideration of further contingencies.

Allowances are defined as costs added to individual estimated costs to compensate for a known shortfall in data, or to provide for anticipated developments. On some projects they can form a major component of the net cost. When placing enquiries for major equipment items within the tender period, the time constraints inevitably result in the enquiries being less detailed than when orders are being placed after the award of the contract. At the tender stage, detailed evaluation and clarifications with the vendors are unlikely to take place and the contractor will be under pressure to quote competitive prices in order to maximize his chances of success. The review of the technical and commercial estimates for the tender will normally result in the identification of additional costs to reflect the anticipated purchase price for these items. In general terms, the more detail that is used to generate the cost, the more consideration is required with respect to allowances.

Contingency is an adjustment to the net cost and is considered no matter how much detailed work has been completed during the preparation of the estimate. The estimate is an estimate, not a detailed costing, and as such will undoubtedly contain uncertainties which will affect the final cost. Contingency adjustments allow for the unknown element and also for any factors outside the control of the contractor which are perceived as likely to affect the final cost.

The objective of the contractor is to win the contract. To achieve this he must submit a technically acceptable bid that generally must also be the lowest priced bid. The process of minimizing risk by adding allowances and contingencies will raise the tender price—eventually to a point at which the tender becomes uncompetitive. However, the contractor is in business to make a profit and alternative strategies can be adopted which have the effect of lowering the tender price but may complicate the task of the client in the evaluation of tender bids. Primarily, these strategies involve qualified bids and alternative bids.

Qualified and alternative bids

Placing qualifications on a bid means that both technical and commercial adjustments by the client may be necessary before meaningful comparisons can be made with conforming tenders on a like for like basis. Assumptions have to be made during the estimating process and the implications of these assumptions may need to be stated formally at the tender stage as bid qualifications. Typically, qualifications can be stated with respect to process conditions, site conditions and the control procedure. They may also apply to pre-commissioning and commissioning activities which limit the engineering and material costs of the project, or have direct cost implications; sometimes they involve transferring costs back to the client. Often some of these limitations will not be acceptable to the client and may have to be added back into the tender at a later stage in the bidding process.

Alternatively, the contractor may decide to price the client's option and also to offer an alternative, lower cost option based on a different process design. However, if the client has adopted a competitive tendering policy, this is likely to preclude discussions on alternative process or engineering options before tenders are submitted.

Award

The contractor with the most acceptable tender submission—which is often the lowest priced bid—is invited to a pre-award meeting, and the final contract award and contract price are then determined by negotiation. The possibility that the negotiated price will be identical to the total estimated price of the project is remote. The more the price is driven down by the client the more the contractor has to respond during the contract to try to recoup elements of profit, contingency or actual cost surrendered at the bid stage. The very nature of the fixed price contract draws a line between the client and the contractor and a potential adversarial relationship develops, with each side manoeuvring for advantage. The client has the advantage during the bidding process, but once he is committed to a contract the advantage passes to the contractor.

Partnership estimating

One new development which is growing in its application is designed to remove the adversarial nature of the traditional fixed price tender. In an attempt to reduce disputes and conflicts in contracts many clients have begun to form long-term relationships with a smaller number of contractors.

The client approaches a contractor during the earliest stages of a project and the contractor is engaged to work with the client during the feasibility and development stages. This enables the client to have early access to specialist expertise to determine the viability of the project. The contractor is paid on a fee reimbursable basis for the estimate preparation. If the client decides to proceed with the project, the same contractor is awarded the contract on the basis of a lump sum, fixed price derived from the completed estimate.

Another alternative is to bid a fixed price for services only, whereby the engineering contractor agrees to provide his design, procurement and management services for a lump sum price.

Materials, equipment and installation contracts are then reimbursed at cost by the client. In this way the client can obtain competitive quotes from engineering contractors on a like for like basis and still retain the benefit of having to pay only the market price for materials, equipment and installation costs.

This form of contract has the advantage of effectively making the engineering contractor an extension of the client's project arm, and in theory should lead to a better unison between the contractor and the client.

Summary

Estimating for construction projects in the process industry has many differences from that for conventional civil engineering and infrastructure projects. The procedure involves the preparation of the process design and estimate in addition to the estimate for the construction, erection and installation works. It is therefore more complex and more expensive although tender periods are not significantly longer. A number of key features at each stage have been identified which, although not exhaustive, should prove useful to engineers tackling this problem for the first time.

The innovation in contractual arrangements of facilitating an open working relationship between the client and the contractor would seem to benefit both parties in the long term. Indeed, the concept of the estimate prepared in partnership would appear to remove many of the main problems of fixed price tender estimating.

Bibliography

Association of Cost Engineers. *Cost engineering technology*. ACE, London, 1987.

Association of Cost Engineers. *Estimating checklist for capital projects*, 2nd edn. ACE, London, 1991.

Association of Cost Engineers. *Standard method of measurement for industrial engineering construction*. ACE, London, 1984.

National Economic Development Office. *Guidelines for the management of major projects in the process industries*, 2nd edn. NEDO, London, 1991.

6 Tender adjudication

This chapter describes both the tender adjudication where the contractor finalizes the bid for the construction work on a project and the client's evaluation of the tenders received.

Finalization of the bidder's tender

The contractor preparing the bid will arrange a tender adjudication meeting before the bid is submitted. The purpose of this meeting is to examine the base estimate and to determine the mark-up, and the allowances for overheads, profit and risk. The agreement on these allowances requires careful consideration of the current market and the need to secure the construction work.

Preparing reports for the bidder's tender adjudication

The tender adjudication meeting will be attended by those responsible for determining the bid price, those persons who have played a significant part in the production of the estimate and representatives of the senior management of the company. Where appropriate the estimator will prepare a set of reports on the base estimate for presentation at the tender adjudication meeting. These reports allow senior management to decide on the appropriate level of mark-up.

These reports should contain

- a brief description of the project
- a description of the method of construction
- a list of problem areas or risks associated with the projects not adequately covered by the contract documents
- any non-standard contract details
- an assessment of the state of the design process
- any assumptions made in the preparation of the estimate

- an assessment of the profitability of the works
- any pertinent information concerning market and industrial conditions
- a review of any relevant past project undertaken for the same client.

In addition the estimator should prepare detailed cost reports which itemize the quantities and costs of labour, materials, plant, subcontractors, nominated subcontractors and suppliers, provisional sums, dayworks, contingencies, attendance on domestic and nominated subcontractors, and amounts included for material and subcontractor discounts. The estimator may also calculate a projected cash flow for the project based on a range of likely mark-ups (see McCaffer and Baldwin, Bibliography).

The purpose of these reports is to convey the details of the contract and the assumptions made in the base estimate to the manager who has the responsibility of deciding the final bid submitted for the work. For large or risky projects this responsibility is usually that of the managing director of the company.

The adjudication process

At the meeting, commercial implications will be considered and additions and allowances decided. These include

- adjustments to the direct cost estimate
- additions for head office overheads
- additions for profit
- assessment of, and additions for, risk
- allowance for inflation
- adjustments for rate loading or bid unbalancing
- assessment of the project cash flow.

Adjustments to the base estimate

The estimators will have prepared the proposed construction programme for the project which will include the basic assumptions and decisions implicit in the estimate, with regard to

- labour, plant and material requirements
- plant and labour output rates
- wastage assumptions

- site overhead requirements.

The senior management should satisfy themselves as to the adequacy of the estimate by interrogating the estimator.

Decisions made by those attending the tender adjudication meeting will include any adjustments required in the resource levels of the project and output or wastage levels.

Additions for head office overheads

Each project undertaken must contribute to the cost of maintaining the head office and general overheads of the company. The financial director will be required to submit the latest details of this cost. In order to make adequate provision for this cost the company must monitor both costs and turnover and know the relationship between these two factors. A monthly comparison of the level of turnover and overhead costs should be kept to indicate the contribution required from future contracts.

Additions for profit

The minimum profit required will be determined by how the profit will be divided. The company will have to face the following demands on the profit it makes

- the payment of corporation tax
- the dividend payable to the shareholders of the equity capital
- the profits that will need to be retained for reinvestment.

Standard corporation tax will be varied by the Government at the time of their annual budget or at the time of special budget statements. The interest on borrowed capital will vary with the rise or fall in interest rates. The dividend payable to equity shareholders will be set by the board of directors with due consideration of the level of dividend required to satisfy shareholders and maintain the share value of the company. The level of profits required for reinvestment will depend on the programme of growth set by the directors and the current level of borrowing from other sources.

The critical financial ratios that determine the company's well-being are

$$\frac{\text{profit}}{\text{turnover}} \text{ and } \frac{\text{turnover}}{\text{capital}}$$

During most of the 1970s and 1980s the profit/turnover ratio experienced by UK construction companies was very low—typically about 0·03. The turnover/capital ratio, however, was relatively high—of the order of 8–15, with 10 being a typical figure. This high turnover/capital ratio reflected the ability to run a construction company on a low capital base. Little capital is required because companies can hire plant, rent offices and receive monthly interim payments from clients.

The combination of these two ratios gives the return on capital

$$\frac{\text{profit}}{\text{turnover}} \times \frac{\text{turnover}}{\text{capital}} = \frac{\text{profit}}{\text{capital}}$$

Thus a profit/turnover ratio of 0·03 and a turnover/capital ratio of 10 give a return on capital of 30%, and perhaps viability. Before servicing debts, contractors would be looking for a return on capital of 30–50%. Thus the following simplified example would hold.

Capital employed (equity £75 000, loans £25 000)	£100 000
Planned turnover ratio	10
Corporation tax	33%
Interest rate	15%
Planned return to shareholders	£7500
Planned return on capital (pre-tax and debt servicing)	30%
Therefore: planned turnover	£1000 000
planned return	£30 000
Interest charges (£25 000 at 15%)	£3750.00
Corporation tax (£30 000 – £3750 at 33%)	£8662.50
Shareholders' return	£7500.00
Total	£19 912.50
Retained profits	£10 087.50

$$\frac{\text{Profit}}{\text{Turnover}} = \frac{£30\,000}{£1000\,000} = 0{\cdot}03$$

The major adjustments to these figures arise from turnover increasing or decreasing with respect to capital employed and decisions regarding the amount of profit to be retained and the return to the equity investors. Thus if the market forces suggested a smaller margin, to be healthy the company would need to increase its turnover with respect to its capital employed, or reduce its retained profits, thereby allowing the company to decrease the returns to the equity shareholders. In a buoyant market, with high levels of both on-going work and anticipated orders, the profit margin would be raised and a much happier scenario would emerge.

Additions for risk

The risk associated with the project must be identified and quantified commercially. This is particularly important on large projects and where the work is abroad. This will involve the estimator identifying potential construction problems with alternatives, areas of the project where insufficient information is available, and item quantities in the bill which appear unrealistic.

Complex mathematical models, including those based on simulation techniques, may be developed for the assessment of risk but in practice these are rarely used by the contractor because of the time normally allowed for the preparation of the bid. The contractor will rely on a simple assessment of the risk elements (e.g. climatic factors, labour availability and materials availability) and the risk categories (e.g. contractual risk, client risk, construction risk and economic risk).

Risk may be offset in a number of ways. Additional insurance may be secured on any relevant item in the construction. It may, for example, be decided to subcontract the work. Moneys must be included within the bid to reflect this additional cost to the contract. Allowance for the risk may be made in the form of a single sum of money equal to an agreed percentage of the base cost of the works. This sum is then added to the profit allowance and the risk treated as a risk margin. If construction proceeds without any problems then a greater profit is achieved. If an unexpected occurrence occurs the cost should hopefully be covered and the profit margin protected.

Allowance for inflation

Depending on the client and the prevailing level of inflation, the contract may contain price adjustment clauses using agreed published indices and the appropriate formula. In times of high inflation contractors will take particular care to satisfy themselves that they are not overexposed to inflation.

Assessment of project cash flow

A project cash flow will normally have been produced during the estimating process. This will be examined to determine whether or not it fits into the overall company cash flow requirements. Adjustments may be made to improve the project's cash flow and minimize the contractor's capital locked up in the contract, and thereby to minimize the contractor's risk. This may be achieved by some form of rate loading.

Adjustments for rate loading or bid unbalancing

Rate loading is the process of adjusting rates throughout the bill of quantities while keeping the total constant. The purpose of rate loading is to minimize the capital locked up and to optimize cash flow without compromising the competitiveness of the tender. Rate loading may be undertaken to obtain greater early income by increasing the rates of the initial site activities (this is known as front-end loading). It may be undertaken to increase profit by increasing the rates on items the quantities of which are believed to have been underestimated and decreasing the rates on items which are believed to be overestimated.

Redistributing rates around the bill manually is a time-consuming and laborious process. However, the advent of computer-aided systems has greatly enhanced the contractor's ability to manipulate rates and fine tune the tender to his own requirements.

Writing up the bill of quantities

Having considered the various factors, and having quantified them, the estimator must finally adjust the bill of quantities (if it is a bill of quantities-type contract) by the amounts decided in the adjudication meeting. This may be achieved by

- increasing all bill item rates by a single percentage
- including a lump sum addition as an adjustment item
- a combination of both.

Usually the time available between the tender adjudication meeting and the submission of the tender is limited and therefore estimators often include a notional percentage mark-up to the bill items at the estimate stage and include lump sum adjustments to implement the decisions made at the adjudication meeting. The introduction of computer systems has enabled these final adjustments to be made faster and more comprehensively than was previously possible.

The client's evaluation of the tender

The contractor will be required to deliver the tender by a stated time, on a stated date, to a specified address. Any tender received after that deadline will be rejected and returned unopened. The tender submission will normally comprise

- the form of tender
- a priced bill of quantities.

In some instances the contractor may be required to submit a bid bond to the client guaranteeing that in the event of

- the contractor's withdrawal of the tendering during the tender assessment period
- failure or refusal to execute the contract form when it is requested
- failure to provide a performance bond

the stated sum will be paid to the client by the guarantor of the bond.

The form of tender will be checked by the client's representative to ensure that the document has been completed properly and has been signed by a person legally entitled to represent the company.

The priced bill of quantities will be scrutinized to identify any omissions, mistakes or exceptionally high or low rates. Although the client's representative will not know how the tenderer has made up his rate for each item, he will be able to compare the bill item rates with those in the other tenders. If a rate within the bill is exceptionally low this may be due to an error. If a rate is particularly high then the contractor may consider that an undermeasurement of the bill item has been made and more work may be available at the time of construction.

It is therefore normal for the client's representative to tabulate the rates supplied by all the tenderers and undertake a detailed comparison. These details will then be supplied to the client, or promoter, in a formal report together with

- a tabular statement of the salient features of all the tenders received
- a statement of any arithmetical errors
- a reference to any discussions between the parties concerned with respect to ambiguities within the tender
- a concise summary of the examination and analysis of each tender
- a comparison of the recommended tender sum with the client's representative's estimate
- a recommendation of the most acceptable tender
- recommendations for dealing with any errors, qualifications and so on with regard to the recommended tender before acceptance
- a financial statement indicating the order of the funds which will need to be made available to meet the payments due on the project.

In the evaluation of the tenders submitted the client's representative will be aware that the lowest bid submitted will not necessarily result in the lowest final price for the construction work. Therefore the lowest bid may need to be rejected because it is unrealistically low. If contractors are bidding for a high-risk contract, the client should consider setting a minimum acceptable risk contract price below which the bidder will not be awarded the contract. Whatever price is chosen, the client should be clear in his mind as to what the level of risk is, and he should be satisfied that the contract has not been awarded to a contractor taking an unreasonable risk.

For the contractor, the period after submission of the tender is one of awaiting the results of the submission. All formal and informal communication channels will be explored to gain feedback on the likelihood of the contract being awarded to his company.

Bibliography

Hillebrandt P. and Cannon. J. *The modern construction firm*. Maxwell Macmillan, Basingstoke, 1990.

Inyang E. D. and Willmer G. Practical considerations in the solution of probabilistic risk analysis models for engineering project management decision making. *Proceedings of international conference on advances in engineering management, Swansea*, pp. 45–60. Johnson, Redruth, 1986.

McCaffer R. and Baldwin A. N. *Estimating and tendering for civil engineering works*, 2nd edn. BSP, Oxford, 1991.

Thompson P. A. and Perry J. G. (eds). *Engineering construction risks —a guide to project risk analysis and risk management*. Thomas Telford, London, 1992.

7 Case study: Cheddleton sewage treatment works

Cheddleton is a large village with a population of about 6500 lying in North Staffordshire on the edge of the Peak District and about eight miles north-east of Stoke-on-Trent. The sewage treatment works lies in a narrow valley adjacent to the Caldon Canal. Its effluent is discharged via a syphon beneath the canal into the River Churnet which is a tributary of the Rive Dove and subsequently the River Trent.

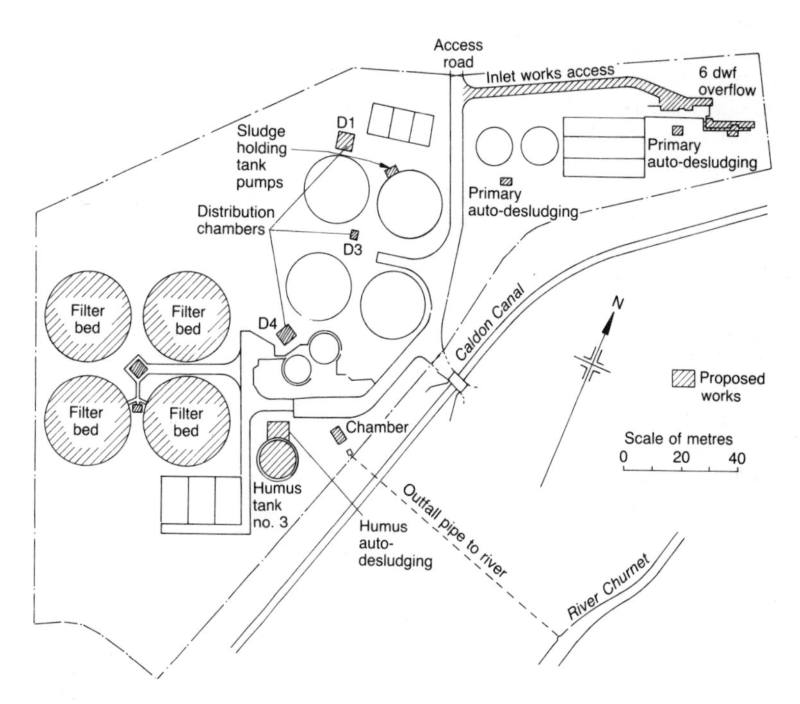

Fig. 6. Plan of Cheddleton sewage treatment works

The quality of effluent from the works was deteriorating some years ago and the works regularly failed to meet its discharge consent. The works was overloaded hydraulically and biologically and extensions were required urgently to enable the effluent to meet the requirement of the Control of Pollution Act and the long-

Table 3. Feasibility estimates (price base March 1987)

	Additional biological filters: £	Oxidation ditch: £	Rotating biological contactors: £
Phase 1. Capital cost			
Biological treatment			
Civil	314 000	190 000	167 400
Mechanical and electrical	61 700	101 500	247 500
Other work			
Civil	89 000	89 000	89 000
Mechanical and electrical	64 300	64 300	64 300
Design and supervision	41 000	30 000	58 000
Land compensation	8 000	8 000	8 000
Site investigation	5 000	5 000	5 000
Total	583 000	488 000	640 000
Revenue consequences, increase per annum over current costs	6 000	25 000	15 000
Phase 2			
Biological treatment			
Civil	100 000	36 000	282 000
Mechanical and electrical	13 000	—	470 000
Design and supervision	13 000	11 000	25 000
Total	126 000	47 000	777 000
Revenue consequences	1 000	4 000	21 000
Total capital	709 000	535 000	1417 000
Total revenue	7 000	29 000	36 000

term objectives of improving water quality in the River Churnet. Planning for remedial action was started in 1985.

Figure 6 shows the general arrangement of the old works together with the proposed extensions.

Feasibility study

Work at the feasibility study stage was concentrated on identification in detail of why the works was failing and which parts of the process required improvment and/or augmentation. The major single problem was a lack of biological treatment capacity and the options to rectify this were identified. The possibility of closing the works altogether and pumping the flow to the much larger sewage treatment works at Leek about two miles away was also investigated, but this was rejected.

The options selected for conceptual design and cost estimating were

- to provide additional biological filters
- to provide an oxidation ditch
- to provide rotating biological contactors.

It was also decided to investigate the possibility of carrying out the work in two phases, phase 1 being designed to resolve immediate problems and phase 2 being designed to meet the long-term objectives.

Common to all these options, other essential work included major improvement to the works inlet and stormwater overflow provisions, including the installation of new screening plant and the construction of vehicular access to this part of the site. The hydraulic bottleneck between sedimentation tanks and filters was to be eliminated and the works pumping station modernized. Also, access to the site, which was along half a mile of single lane farm track, was to be improved.

Feasibility estimates

Feasibility estimates were built up using TR61,* manufacturers' budget quotations and information derived from tenders for similar schemes. TR61 contains cost estimating models for a wide range

---

* Water Research Centre. *Cost information for water supply and sewage disposal.* WRC, Henley, 1977, TR61.

of works and works components which use historic information in conjunction with inflation indices to give broad-brush estimates. For Cheddleton TR61 was used to estimate the cost of the individual components of the extensions (e.g. the biological filters) but an assumption had to be made at this stage about the extent of other improvements and modifications which it might prove necessary to incorporate into the existing works. The estimated figures are given in Table 3.

Conceptual design and costing

The feasibility study revealed a significant cost penalty in respect of the option to provide rotating biological contactors and initially some work was done to check the validity of the design assumptions. The assumptions were verified and at an early stage this option was eliminated. A check was also made on the feasibility of constructing an oxidation ditch in two phases, as a result of which it was resolved to drop this approach.

During the conceptual design process further work was carried out on the feasibility of constructing a new access road to the site, and more detailed work was carried out on the siting, levels, general arrangement and size of the proposed structures and pipelines. Some additional work was also identified.

Estimates for the remaining options were then built up using the additional information available, but costings were still dependent on TR61, manufacturers' budget quotations and information derived from tenders for similar schemes. The figures estimated at this stage are shown in Table 4. The options for comparison were

- to construct biological filters in two phases
- to construct biological filters in one stage
- to provide an oxidation ditch.

Detailed design and costing

On the basis of the information available from the conceptual design stage, it was decided to proceed with detailed design for the provision of additional biological filters in one stage to achieve the long-term objectives. Major practical difficulties were identified over the construction of a new access road and this part of the scheme was abandoned.

Table 4. Concept design estimates (price base January 1988)

	Construct biological filters in two phases: £		Construct biological filters in one stage: £	Oxidation ditch: £
	Phase 1	Phase 2		
Biological treatment				
Civil	157 300	95 000	586 000	495 000
Mechanical and electrical	61 000	9 000	70 000	101 000
Other work				
Civil	172 700		172 000	172 700
Mechanical and electrical	193 000		193 000	193 000
Design and supervision	100 000	15 000	111 000	83 000
Land compensation	4 000		4 000	4 000
Site investigation	5 000		5 000	5 000
Total	1053 000	119 000	1142 000	1054 000
Revenue consequences per annum	8 000	9 000	9 000	22 000

Initially a thorough survey of the site was undertaken to confirm the validity of existing drawings and to establish the extent of underground apparatus and pipelines. A contract was let for ground investigation and discussions were started with staff responsible for operating the works in order to establish in detail how the extensions and improvement could be carried out while sewage treatment on the site was maintained.

The ground investigation confirmed that there was a high water-table on the site which would necessitate dewatering techniques during the construction and permanent anti-flotation measures for some of the new structures. The survey confirmed the existence of many underground pipelines on the site and the difficulties of negotiating these in confined areas were evaluated.

It was decided to let a single civil engineering contract and several mechanical and electrical contracts for the different types of plant and equipment required. A detailed design and drawings were produced, followed by bills of quantities. Contract documents, including extensive requirements in the specification relating to

Table 5. Detailed design estimate mid contract (price base December 1989)

Description	Cost: £
Biological treatment	
Civil	1134 148
Mechanical and electrical	140 000
Other work	
Civil	378 958
Mechanical and electrical	434 000
Design and supervision	176 000
Land compensation	6 000
Site investigation	6 000
Total	2275 106

Table 6. Accepted tender (price base June 1989)

Description	Estimate: £
Civil contract	1407 000
Mechanical and electrical contracts	492 000
Design and supervision	191 000
Other costs	26 000
Total	2116 000

Table 7. Final costs

Description	Out-turn cost: £
Civil contract	1463 000
Mechanical and electrical contracts	526 000
Design and supervision	216 000
Other costs	31 000
Total	2236 000

Table 8. Estimates at each stage of the scheme

Description	Stage 1: feasibility estimate (price base March 1987)	Stage 2: concept design estimate (price base January 1988)	Stage 3: detailed design estimate mid contract (price base December 1989)	Stage 4: accepted tender (June 1989)	Stage 5: Final costs
Civil contract	503 000	758 300	1513 106	1407 000	1463 000
Mechanical and electrical contracts	139 000	263 000	574 000	492 000	526 000
Design and supervision	54 000	111 000	176 000	191 000	216 000
Other costs	13 000	9 000	12 000	26 000	31 000
Total	709 000	1142 000	2275 106	2116 000	2236 000
Inflation indices (as published April 1992)	115·1	123·5	155·3	155·3	148·7

the method of construction and sequencing of works, were then produced. Finally, a detailed estimate based on the bills of quantities was produced. Prices were drawn from up to date tenders for similar works. The estimate produced is shown in Table 5.

Tender and construction stage

Competitive fixed price tenders were invited for each contract and a pre-construction stage estimate was then produced based on the successful tenders. This estimate is summarized in Table 6.

Construction commenced in June 1989 and the works were substantially completed 12 months later. The sewage works then operated satisfactorily and within its consent.

Final costs

Final accounts were settled without major difficulty. The final costs of the scheme are shown in Table 7.

Discussion

The estimates produced at each stage of the scheme for the chosen option and the final cost are shown in Table 8.

Making adjustments for inflation, the estimated cost of the scheme rose 50% between stages 1 and 2 and 58% between stages 2 and 3. After stage 3 the scheme progressed through the construction stage with a minimum of unforeseen problems; this is reflected in the fact that the out-turn cost was very similar to that estimated at stage 3.

Between stages 1 and 2 extra work was added to the scheme by the addition of a sludge tanker loading facility. This had both civil and mechanical and electrical cost implications and some mechanical and electrical plant renewal was also included. By the time the stage 2 estimate was produced an inspection of the site had taken place and the difficulties of carrying out the work on a rather isolated and confined site were being recognized. The civil costs rose 40% and the mechanical and electrical costs by 76%.

Between stages 2 and 3 a detailed topographical survey and ground investigation identified many problems, including a high groundwater level. The existing site was more congested with underground cables and pipes than had been originally anticipated and many practical problems were identified which would affect the sequence of construction and so on in order to keep the existing works operational throughout the construction period. All these issues influenced the detailed civil engineering design and the contract documents which were produced. When the documents were priced to produce the stage 3 estimate it was known that civil engineering costs generally were rising rapidly. The estimate at this stage for the mechanical and electrical work was also heavily influenced by extraordinary increases in sewage works plant costs—the level of engineering activity in the water industry was rising rapidly at the time with the onset of privatization, and demand was tending to outstrip supply.

Findings

This case study illustrates some key issues which affect estimating

- the difficulty of estimating accurately at an early stage in scheme development where the works are essentially modifications and extensions to an existing site—this situation is very different from greenfield conditions
- the problems of using historical data for estimating
- the speed with which price trends can suddenly change through market forces; in the two years following the remedial work done at Cheddleton, prices in the water industry fell dramatically, reflecting the general downturn in engineering activity and the depressed economy
- the fact that the accuracy of an estimate depends greatly on the effort which goes into producing it and the extent to which the overall scope of the scheme has been properly identified at the time.

Part 2. Estimating practice

This part is concerned with information technology (IT) and the human factors of estimators, and considers their influence on estimating.

The use of existing IT systems and the development of new approaches are examined and problems of data capture and reliability are reviewed. Database and non-database programs are compared but specific commercial programs are not identified. Developments in estimating, including modification to bills of quantities into cost-significant work packages, and the linking of planning and estimating software are also discussed.

In identifying human factors in estimating reference is made to research conducted in the UK into the influence of the experience, the role and the skills of the estimator and in the interpretation of data. The human element in estimating has always been recognized as being significant but has rarely been addressed in a practical manner.

8 Information technology in estimating

Recent changes and developments in IT have influenced estimating activities significantly in two ways. First, software packages following traditional estimating procedures are now available for data manipulation and processing. Second, new methods of estimating have developed that are encouraged by the application of computers. The trend seems to be to move away from using IT for data capture, manipulation and retrieval in massive databases and towards its use in less mathematically rigorous methods of analysis to tackle estimating problems which usually contain uncertainty.

This chapter reviews the basic strengths and weaknesses of adopting IT in civil engineering estimating and considers the viability of using common data for the development of the appraisal estimate, the detailed estimate and ultimately the project management system for the construction of the works. Areas in which IT might be developed for estimating, including the parametric estimating approach and the use of knowledge-based systems, are reviewed briefly.

Computer-aided estimating

The principal functions of computer-based estimating techniques are to facilitate data retrieval and transfer, to perform data manipulation, usually in the form of rapid calculation, and to produce reports. As in manual estimating, the role of computer estimating will depend on the user's requirements. Those of a client at the design stage will be very different from those of a contractor at tender. The data available for the production of an estimate will also be different.

There are certain basic functions that computer software must incorporate whatever method of estimating is to be used

- a data library for storing item or resource data
- a range of methods for adding to the direct costs to produce a price
- the ability to update or alter any of the input data and to recalculate the estimate
- reporting facilities to enable the estimator to produce details and summaries of the estimate and input data.

To be user-friendly it is important that software performs the basic functions and allows data input or price build-up by the estimator in his normal way of working. The estimator's methods and accuracies may be refined as he has to spend less time performing routine calculations and arithmetical checks, but his basic methods must remain unchanged or his judgement may be affected.

A computer-based estimate is only as accurate as the input data; the use of a computer in itself will not necessarily increase the accuracy of estimates. Arithmetical errors should disappear but if the database is not kept up to date or if it is applied without careful thought then a decrease in accuracy could result.

Estimating data

Some basic questions concerning the data should be considered before the appropriate method of estimating is decided. Typical questions include the following.

- Where do the original data come from?
- Can data be imported/exported to other software packages?
- How accurately do the data relate to the geography and history of the site?
- What format should the database take (e.g. by resource with labour and plant rates per hour, output, on-costs, material rate per unit quantity, wastage, common descriptions of packages/items, unit rates)?

Advantages and disadvantages of computer utilization

Advantages of computer utilization are

- data utilization
  - communication between departments
  - ease of data access and retrieval

  - o data manipulation (alternative designs)
  - o common assumptions from job to job
- control
  - o cash flow (effect of loading strategies)
  - o buying
  - o cost monitoring (cost centres)
  - o resource utilization
- reports
  - o clear and in different formats
  - o summaries
- errors
  - o fewer calculation errors
  - o fewer misreading errors
- rapid, almost universal, data transfer.

Disadvantages of computer utilization are

- changes in established practice
  - o sources of error
  - o training implications
  - o possible confusion of responsibilities
- tender procedures
  - o submission still has to be handwritten
  - o few invoicing systems are computerized
- accuracy
  - o use of historic data
  - o meaning and value of output.

Client's estimate

Traditionally, if the estimator worked for a client he would price the items from rates tendered for other similar projects, as it would be unlikely that he would have sufficient knowledge of the construction process to produce an operational estimate. The purpose of this estimate would be twofold: to gain an approximate idea of how much the job would cost and for use as a basis for comparison with contractor's bids, in an attempt to discover any suspiciously high or low rates.

The pricing of a full bill of quantities may not be viable, due to insufficient development of the design or the lack of price data, and other estimating options may have to be considered. Typically, an approximate bill may be priced, with only the major elements of the project priced and an allowance made for the others, or historic unit rates may be used. Computer packages are available to aid the estimator in these cases.

Contractor's estimate

The contractor's estimator has to decide on the amount of plant and labour to use for a certain operation. The rate then depends on the assumed output rate, the source and quantity of materials which will be used, the costs of setting up the operation and of the temporary works, and the contribution which the item is to make to indirect costs, overheads and profit. The translation of this operational cost into a unit rate confuses control and tends to make valuation adjustments unrealistic. It does, however, give clients an equivalent basis for comparison.

A number of computer packages are available to help the contractor with this task. The methodology should closely reflect the manual estimating procedure. The removal of routine calculations frees the estimator to concentrate on pricing policy and optimization of the tender. Common software packages include

- spreadsheet systems based on rate build-ups
- bespoke systems developed specifically for use within one organization (these necessitate a very expensive and long development period)
- estimating packages of specialist software.

Integration of time and cost

Most up to date estimating software packages can be linked with other computer programs. In particular, to rationalize measurement and valuation and to produce a systematic method for the valuation of variations, delays and disruptions by linking time and money, the estimating package can be linked with a planning package, and data transferred automatically between the packages.

The development of an integrated time and cost system has been inhibited by the problem of allocating a multitude of bill items to a lesser number of activities. This problem is not very significant at a more general programme level and will become even less significant if, as the result of current research, the number of items in the bill of quantities is reduced.

Through the automatic transference of the tender build-up data, including an estimated number of hours of use of each resource, the resource cost per hour and the allocation of resources to activities, the planner and estimator can reconcile their estimates of time and money quickly and easily. Planning time is saved as there are many fewer data to be input. Through levelling of the critical resources, overheads can be reduced, producing a much more competitive bid.

Planning decisions, such as network logic and the availability of resources, are left to the planner. It is up to him, through resource levelling and scheduling, to optimize the programme to balance the need for a short contract duration with the demand on resources. Any resultant savings should be discussed with the estimator before deciding to adjust the bill of quantities. The planner cannot decide to do this as the prices associated with individual items do not necessarily reflect the cost of carrying out that item of work; the adjustment is not therefore straightforward and could take a variety of forms.

Not only should the estimating and planning departments become integrated through the use of a computer system, but also through computers the work of all departments can be integrated, so that problems of work duplication and easing communication are overcome, giving the manager a clear project overview and minimizing the number of subjective decisions he has to make.

Cost-significant work packages

Research has shown that typically 20% of the items in the bill of quantities represent 80% of the value. In other words 80% of the bill items are individually insignificant. They are time-consuming to price and make linking time and money almost impossible. This link is very important if variations, delays and disruptions are to be evaluated fairly as most of a contractor's costs are time-related. If the insignificant items could be identified and eliminated from the bill of quantities, which would then be known as a 'simplified bill', the estimator's task would be simplified.

An item is insignificant if it is

- of less than the mean value (i.e. of less than the contract sum (minus preliminaries divided by the number of items) (cost insignificant))

- in the lower 70% of bill items in terms of quantity of each section of a bill (quantity insignificant)
- of less than the mean value in terms of the number of hours required for plant and labour usage (resource insignificant).

Once the items of small value have been eliminated, the remaining significant items can be packaged with similar items without any adverse effect on accuracy. The only requirement is that the work must be executed by one type of gang which is associated with a specific material and one productivity output. The remaining groupings of items are known as 'work packages'. A standard set of work packages holds a fixed proportion of the contract sum for a certain type of job.

If, for example, a reinforced concrete bridge is considered, the work packages would be classed under the categories of end supports, intermediate supports, superstructure and finishes. The work packages in the end support category would be

- drainage of substructure (sum)
- excavation in any material other than rock or artificial hard material in or adjacent to foundations; any depth (m^3)
- excavate rock in or adjacent to foundations; any depth (m^3)
- disposal of excavated material off site (m^3)
- filling in, or adjacent to, foundations; any depth; any material (m^3)
- formwork >300 mm wide; 85–90° to horizontal; any class other than patterned or curved (m^2)
- reinforcement; any diameter; any length of bar; mild steel or high-yield steel (t)
- concrete any mix, excluding blinding (m^3)
- concrete protection or any waterproof coating applied to concrete surface (m^2).

The actual proportion of the total project value that the work packages hold can be determined; it is usually 80–90%. This proportion is different, but constant, for the various categories of project, such as roads, bridges and water treatment works. This proportion directly translates into the cost model factor, which can be used by estimators to obtain quickly and accurately the cost of a project. If the cost model factor for reinforced concrete bridges were 0·85, the estimator would divide the total value of the work packages by 0·85, and thus obtain the total contract sum.

The estimator prices the fixed, time-related and quantity-proportional charges associated with each package, specified separately for measurement and valuation purposes. As there are a small number of packages for each job the pricing of three elements rather than one is not significantly more time-consuming. Plant and labour costs are assigned to time charges, materials to quantity-proportional charges, and start up and finish costs to the fixed charge. The use of general method related charges is encouraged.

The importance of the link between money and time cannot be overstressed. The estimator must consider the interaction of the two things in order to optimize his bid, as the efficiency and interaction of resources influences the competitiveness of the bid. The estimator must also consider the cash flow implications of his bid; this takes time to derive from a traditional bill of quantities and programme but is obvious when the payment mechanism is directly linked to a meaningful programme, particularly if method-related charges have been used.

The benefits and limitations for estimators of reducing the number of bill items include the following.

- The linking of money and time is more straightforward and so leads to quicker and more equable evaluation of delay and disruption at the tender stage and as the job progresses; it also saves administrative time.
- Pricing risks are reduced because it is easier to check on the actual cost of the work packages as work progresses on site and to feed these back to the estimator.
- Tendered rates for the small value, or insignificant, items are often inaccurate.
- It is easier to establish a database of reliable subcontractors and materials' suppliers for each work package, speeding estimating as the system is established.
- The tender make-up is more accessible. Hence the tendered rates have to be realistic as loading would be obvious, the information would be easier to use in the event of a claim, cost

comparisons would be easier and the information could be used by clients to predict future project costs.

- Testing has shown consistent results in the value of the cost model factor within one work group.
- An overall reduction in the number of items is obtained, although the estimator has to check exactly what is included in each work package.
- As the estimator becomes familiar with the work package philosophy he will recognize what is included in each package and so the whole process should be quicker.

Parametric estimating

Conventional methods of initial estimating require detailed information about a project at a stage when there is little information to work on. This is in conflict with the fact that accurate cost estimates are vital at this stage, particularly because management need to decide whether or not to go ahead with a project. A unique attribute of the parametric method is that it tackles the estimating task by a top–down approach instead of the traditional detailed or bottom–up approach. It is therefore suited to being used at the stage of a project when little information is available.

If the quality of the preliminary estimate could be improved, then the risks associated with the unknown would be reduced, as would the risks associated with underestimates or overestimates. With the advent of computer-aided planning techniques and the recognition of statistical methods, the parametric approach produces acceptable estimates of cost early in the life cycle of the project, enabling realistic initial budgets to be formed as the basis of the project.

Parametric models are based on one fundamental cost principle

$$\text{cost} = f(K, M, T, R)$$

where K is complexity, specification, quality and application, M is mass, size, quantity and number of parts, T is time span and schedule constraints, and R is resource effectiveness, relative simplicity, experience and design inventory (based on PRICE—parametric review of information for costing and evaluation*).

---

* Spix H. The cost estimating problem and project management. *Trans. Nordnet '91, Trondheim*, 1991, 401–426.

Spix defines a parameter as 'a quantity whose value varies with the circumstances of its application'.* In constructing any model the process begins with the retrieval and evaluation of pertinent cost data from a historical database and its comparison with a physical feature. By regression analysis the data can be reduced to a cost-estimating relationship. A parametric cost-estimating model is an aggregation of cost-estimating relationships which relate engineering features to cost. The traditional method is to break down the labour, plant and material cost components and calculate them separately.

Parametric estimating was developed for manufacturing processes where the elements were clearly defined. The construction industry resisted the introduction of systems of this type by arguing that, by its nature, civil engineering was unsuitable for anything but traditional estimating techniques. With the growth in availability of powerful computers, parametric cost models have been used increasingly, so that it is now possible to estimate costs and schedules for the development and production of every aspect of a project from its conception to support through its working life, and the construction industry is no exception.

Knowledge-based approach to estimating

Expert systems are attracting considerable interest as potential aids to decision-making. In an advisory role they could provide the necessary assistance to produce the cheapest cost estimate, by allowing automated decision-making to select resources to match different workloads.

Estimators generally calculate construction costs by analysing appropriate unit rates and applying these to the measured quantities of the proposed works. Computer-aided estimating is established to varying degrees in most large construction companies. The common feature of these systems is that they still rely on the estimator selecting resources, details of which are held in a library; this draws considerably on the user's judgement and experience. The expert approach purports to provide tools for the effective representation and use of knowledge developed from experience, thus enabling the optimization of resource selection.

---

* Spix H. The cost estimating problem and project management. *Trans. Nordnet '91, Trondheim*, 1991, 401–426.

A computer simulation begins with a module in which the user responds to a set of questions posed by the system. These questions concern the details of the contract. The simulation then progresses through a number of modules relating to the works. The user moves on through the system to determine resources and methods, until finally the cost computation module summarizes the results of all of the cost computations. The summary is given under cost headings which include materials, labour and plant for individual activities and for the complete works.

Estimating systems are evolving within an integrated environment of construction management systems. A knowledge-based system should also develop within this environment as a decision-making or decision-suggesting aid.

Although expert systems have their advantages and in a few years will doubtless lead the market, currently they are expensive and cannot truly emulate the decision-making process on the basis of experience. In the present economic climate there are also doubts that any organization would trust a computer to make key commercial decisions in the pricing, not just the costing, of a bill of quantities.

Bibliography

Bower D. *et al.* Integrating project time with cost and price data. In *Developments in civil and construction engineering computing, Proc. 5th Int. Conf. Civ. Struct. Engng Comput.*, 1993, 41–49.

Institution of Civil Engineers. *The New Engineering Contract.* Thomas Telford, London, 1993.

McGowan P. *et al.* The role of integrated cost and time models in conflict resolution. In *Management and resolution, Proc. 1st Int. Conf. Construct. Conflict*, Part 3, 1992, 255–269.

Smith N. J. and Husby O. Application of project management software in construction projects. In *Trans. Nordnet '91, Trondheim*, 1991, 129–138.

9 Human factors in estimating

It is commonly held that the ability of the estimator to apply professional skill and judgement is an important factor in the production of an accurate cost estimate. This chapter identifies the human factors of the individual estimator that affect estimating accuracy.

Factors affecting accuracy

Estimating is not an exact science: it is as much an art as a science, and involves intuition and expert judgement. The following major factors determine the quality of estimates

- the estimating technique used
- the availability of reliable cost and design information
- the type and size of the project
- the extent to which feedback is used
- the estimator himself.

The abilities and attributes of the individual estimator are a significant factor in accurate estimating. They include the following

- his ability to forecast and interpret trends
- his breadth of, and ability to learn from, experience
- his ability to exercise professional judgement
- his knowledge and academic background
- his ability to work intuitively where design information is limited
- his personality, motivation and enthusiasm for estimating.

The most significant of these human factors is the estimator's knowledge of general price levels. Invariably this is acquired through experience.

Role of the estimator

There are essentially two types of estimator working in the construction industry: contractors' estimators and design estimators. They represent the production and design sides of the industry respectively.

The role of the contractor's estimator is predominantly one of forecasting the net cost to the organization of undertaking predetermined construction works within a specified time constraint. It also involves advising higher management of the decision to tender and on tender levels in general.

The estimator is required to analyse the components of construction—the availability, cost and appropriateness of methods of construction (plant and labour), materials and supervision—and to assess anticipated market conditions. As the estimator is attempting to forecast the likely costs of construction, a knowledge of previous costs is essential. Also, as the organization has to be competitive in order to remain in business, the estimator has to balance the anticipated costs of construction with the need to submit the lowest tender to obtain work. In most contracting organizations the decision to tender and the level of mark-up to apply are separate choices made by management based on the advice of the estimator.

The role of the design or pre-tender cost estimator—which in the UK is often carried out by quantity surveyors or cost engineers—is not as easily defined. It is not always clear whether the design estimator is attempting to anticipate the lowest tender or the final sum payable by the client. The basic task of the design estimator is to predict likely future costs to enable decisions to be made concerning the size, type and specification level of the project and its viability. Unlike the contractor's estimator, who is usually provided with relatively comprehensive design information, the design estimator is required to provide estimates at various stages throughout the design process. To cope with these varying degrees of uncertainty surrounding the details of the target project, the design estimator needs to be proficient in the use of a variety of estimating techniques and to have knowledge of and access to the appropriate cost information.

Although the basic roles of the contactor's estimator and the pre-tender cost estimator are similar (i.e. they are both attempting to predict construction prices), the techniques and quality of information available to them vary. However, the design estimator needs to have greater skills in visualizing and anticipating the designer's decisions and the ability to incorporate them intuitively into the estimate.

Skills of the estimator

Desirable qualities in a good estimator are

- a good basic numerate and literate education
- a reasonable time spent on site
- the ability to read and interpret drawings
- the ability to communicate
- the facility to make accurate mathematical calculations
- the application of logic and common sense
- patience
- a sense of humour
- neat, methodical and tidy by habit
- the ability to cope with a vast volume of paper
- a working knowledge of all the relevant trades
- an appreciation of all facets of the business
- curiosity
- confidence
- the ability to question the basis of assumptions
- a close relationship with those responsible for construction
- a knack of picking up useful information
- flexibility.

The estimator's most valuable attribute is experience. This is acquired over a period of time by learning from colleagues and through the correction of mistakes and errors of judgement made on previous projects.

More specifically, the ability to forecast construction costs accurately requires the estimator to have developed knowledge and understanding of

- the rules of measurement relevant to the project
- current and anticipated market conditions
- the needs and requirements of clients and designers
- the interrelationship between the resources of production and components of design
- contractual obligations.

A formal specialized qualification in estimating, although

advantageous, is generally not an essential requirement for a practising estimator.

Characteristics of the estimator

The following characteristics, in order of importance, are acknowledged to be important in good estimators

- those which enable the estimator to understand the nature of the project
- those which enable the estimator to carry out the estimate
- those which enable the estimator to attain these skills and judge his level of expertise.

Research has shown that, in general, estimators are practical, evaluative, realistic, experienced, logical and pragmatic rather than experimenting, reflecting, watching, intuitive, observing and feeling. The best estimators are generally optimistic, and have a high regard for their own estimating ability and the role of work experience. They also have a knowledge of market conditions.

In a recent questionnaire survey, the highest-rated personal qualities of a design estimator were knowledge and confidence; the lowest-rated were pleasantness and toughness. Most estimators considered themselves to be logical and systematic, confident and helpful, while actively seeking more information to avoid making risky decisions.

The characteristics that were considered to contribute to estimating expertise were concerned with judgement, particularly within the context of the forecasting task. Seven major factor groupings have been established

- professional competency
- psychological aspects
- training
- attributional factors
- risk acceptance
- data processing skills
- experience.

Interpretation of data

The role of the contractor's estimator can be divided into those duties which require programmed skills and those which require judgemental skills. Programmed skills are those which are developed through rote learning and are usually associated with the technician level. Judgemental skills are acquired through experience and individual ability and are associated with the professional level.

In order of importance, the judgemental skills needed are

- assessment of tender levels
- project appreciation
- decision to tender
- overhead and risk consideration

and the programmed skills are

- reconciliation of costs
- material enquiries
- subcontractor negotiation
- analysis of resources.

The interpretation of data is often needed, be it the assessment of the letter of invitation to tender, the tender document, historical data concerning overheads and risk consideration or the future trends in the market. This requires the contractor's estimator to exercise highly refined decision-making skills. Also the design estimator has to possess similar skills in the interpretation of client and designer requirements, historical cost data and the current and future trends in the market.

Human decision-making is often assumed to be based on the individual's interpretation of the available data. This interpretation may often be incorrect as the judgement process is affected by the individual's environment (e.g. by events, people and organizational culture) and reaction to it. Unfortunately there is no generally accepted theory or set of rules to explain the psychological processes of the decision-maker. However, an important factor in distinguishing decision-making abilities is perception, e.g. perception of the available decision options and perception of the possible outcomes resulting from each option. Estimators with highly developed perceptual and intuitive skills should be able to interpret the available data accurately and consistently and therefore to produce more reliable estimates. Most of this intuition is experiential and is based on the amount of data assimilated by the estimator.

Influence of experience

All the anecdotal and empirical evidence available points to the fact that experience is by far the most important factor affecting the performance of both contractors and design estimators. This raises the question of what is meant by experience. In practice, experience is often equated to length of service or age. However, research has shown that what is really important is more specific experience, such as estimating projects of a similar type and size. Indeed, this job experience has been found to correlate closely with both accuracy and consistency of estimating performance. Unfortunately, although indicating the general circumstances in which experience may be acquired, this adds little to the knowledge of what constitutes valid experience. This knowledge would hopefully provide insights into the crucial issue of how what it is that is learned from experience may be acquired in an efficient and effective way.

There are two aspects to experience: the actual participation in an activity and the knowledge or learning derived from it. To quote Aldous Huxley: 'Experience is not what happens to you; it is what you do with what happened to you'. Merely to associate experience with a period of time, a number of similar projects or a range of projects is therefore incomplete. Fifteen years' experience as an estimator could well be 15 times one year's experience or even 45 times four months' experience. Therefore, although a continuity of estimating experience is essential in relation to both the time span and the variety or specific task experience, it is the estimator's reaction to the experience that is significant. The issues at stake are how the individual learns and develops from the experience, and the factors that result in a questioning or change in an individual's working methods which ultimately results in proficient estimating.

Learners can make the most of their experiences when they can recognize learning opportunities and possess the characteristics which enable them to learn effectively from them. Individuals can learn from experience through the following opportunities

- at work
  - unplanned learning in their existing posts
  - planned, created learning within the responsibilities of their existing posts

  - o planned, created learning through increase in their responsibilities
  - o planned, created learning by experience outside the working environment
  - o planned learning from supervisors or colleagues

- outside work
  - o courses, seminars, conferences and workshops
  - o voluntary work.

Most learning opportunities fall into the category of unplanned learning within the individual's existing position; few estimators have received any post-qualification training in estimating. Also few organizations have developed any planned learning experiences for staff apart from initial job rotation for trainee staff to provide a basic knowledge of the function and responsibilities of the various divisions or sections within an organization. For most estimators, therefore, learning from experience falls within a purely accidental and unconscious process resulting from the participation in an activity. This process is essentially one of reaction to events—especially unpleasant experiences—and does not necessarily develop the desired attributes.

To benefit from their experiences, learners require the following characteristics that enable them to recognize learning opportunities as they present themselves

- the capability of being dissatisfied with current levels of performance, knowledge, skills or attitudes
- the capability of recognizing that activities can have more than one purpose
- the belief that it is possible to learn by planned direction rather than by accident
- the belief that the culture in which they work, and in particular their employers, will give them some support and some reward
- the belief that recognizing learning opportunities will lead to an improvement which they desire to make.

It is generally accepted that in order to learn more effectively from their experiences, learners should

- review their experience frequently
- openly share their experiences

- respond flexibly to the unexpected
- reach conclusions via careful thought
- have detailed recall
- be able to bridge the gap between artificial situations and reality
- put deliberate effort into learning
- ask questions
- listen patiently
- express thoughts fluently
- be open to new angles and possibilities
- identify their own development needs
- be able to convert ideas into feasible actions
- take risks
- see connections
- ask for feedback
- adjust quickly to unfamiliar situations
- make specific action plans
- convert criticism into constructive suggestions for improvement.

To achieve this, ideal learners in general should refine their specific skills in order to

- establish their own effectiveness criteria
- measure their effectiveness
- identify their own learning needs
- plan their own learning
- use learning opportunities
- revise their own learning processes
- listen to others
- seek out and accept help
- confront any unwelcome information
- take risks and accept uncertainties
- observe the learning of others
- learn about themselves
- share information and receive feedback
- review what has been learned.

There are four ways in which learning from experience may be improved

- increase the amount and immediacy of useful feedback
- create a social environment that requires learning

- hire or train employees to be experts in both substance and process
- do not expect infallibility.

Regarding infallibility, although learning from experience requires one to be wrong at least part of the time, both public and private organizations tend to punish mistakes.

It would appear that people who learn the most through experience accept responsibility for their own development, actively seek feedback, are open to experiences and possess the ability to reflect. The ability to apply critical reflection is a significant factor in learning from experience.

In view of this it is perhaps surprising that research in the field indicates that many estimators tend not to be very open-minded, to be reluctant to change working practices, to use little or no reflective thought, to fail to establish effective feedback systems and fail to learn effectively from experience. The indications are that the development of personality factors through the establishment, the use of effective feedback systems and the ability to learn from experience are the major contributory factors to becoming a proficient estimator.

Bibliography

Ashworth A. *The computer and the estimator*. Chartered Institute of Building, London, 1987, Occasional paper 81.

Ashworth A. and Skitmore R. M. *Accuracy in estimating*. Chartered Institute of Building, London, 1983, Occasional paper 27.

Brandon P. S. *et al*. *Expert systems: strategic planning of construction projects*. Surveyors Publications, London, 1988.

Feldman, J. On the difficulty of learning from experience. In *The thinking organisation* (edited by H. P. Simms). Jossey Bass, San Francisco, 1986.

Flanagan, R. *Tender price and time prediction for construction work*. PhD thesis, University of Aston in Birmingham, 1980.

D. A. Gioia and Associates. *Dynamics of organisational social cognition*. Jossey Bass, San Francisco, 1986.

Honey P. and Mumford A. *The manual of learning opportunities*. Honey and Mumford, Maidenhead, 1989.

Honey P. and Mumford A. *Using your learning styles*. Peter Honey, Maidenhead, 1986.

Lowe D. J. *Experimental learning: a factor in the development of an expert pre-tender estimator*. MSc thesis, University of Salford, 1992.

Mumford A. C. Learning style and learning skills. *J. Mgmt Dev.*, 1982, **1**.

Mumford A. C. *Making experience pay—management success through effective learning*. McGraw-Hill, London, 1980.

Skitmore R. M. *Contract bidding in construction*. Longman, Harlow, 1989.

Skitmore R. M. *et al. The accuracy of construction price forecasts.* University of Salford, 1990.

Smith R. C. *Estimating and tendering for building work*. Longman, Harlow, 1986.

10 Conclusion

The initial view stated in the preface of this guide was that the objective of an estimate was to provide the most realistic prediction of cost and time, no matter at what stage the estimate was undertaken. To realize this objective there is a fundamental need for relevant data. The main problem areas in estimating relate to difficulties with access to data and with the methodology for the manipulation of data, particularly at the early stages of a project. This is important because early estimates tend to shape the perceptions of a project.

This guide shows that, with a few exceptions, there are three main factors which could lead to significant project underestimates. The first is the inappropriate assessment of risk within an estimate; this factor alone could account for significant variations in actual performance. Second is the use of inappropriate contract strategies, and third the human characteristics of the individual estimator. This emphasizes a wider view of estimating, which should take into consideration not just the physical project elements, but also the sources of risk and the methodology used to execute the project.

Given these weaknesses the clear theme which has been identified in this guide is that at any given project stage the most appropriate estimating technique should be used as part of the continuing process of managing the project.

A number of developments in estimating are currently being pursued; of these, four in particular might represent significant progress

- joint definition of the estimate
- activity schedules
- the use of tender mark-ups
- information technology.

Joint definition describes the process whereby the client and the contractor(s) work closely together on a project to derive and agree the estimate for the project jointly and then use the joint estimate in the execution of the project. This procedure has been used on some turnkey projects when time is short and the design brief is incomplete, and it would seem to offer advantages to both parties. Frequently, when using a fixed price, package contract, one of the main sources of waste and dispute is the contractor's estimated cost. However, if the estimate has been determined jointly, with the contractor being awarded an interim amount to continue work, then the wasteful process of preparing two independent estimates is avoided, and when changes or problems occur later in the project the client is confident in the estimate so that the need for detailed and often protracted negotiations for each event is eliminated.

Activity schedules are schedules of the activities comprising the work in the contract, including design and off-site assembly, with a price shown against each one. An activity schedule differs from a bill of quantities in that

- remeasurement of quantities is required only for variations; activities which remain unchanged are paid for on their completion
- the total of the sums associated with the different activities of design, manufacture, supply construction and commissioning makes up the contract sum; variations are valued by altering the lump sum for varied activities and by adding or deleting activities as required; the schedule may include some quantities in order to simplify the valuation of variations
- the activity schedule is prepared by the contractor; it is possible for the client to specify certain elements that must be included, given that the contractor may expand on them.

It is important that both client and contractor are satisfied that the schedule realistically reflects the work to be done, so that payment reflects actual expenditure, provided the work is executed as planned and to programme. The aim is not to increase the workload of the estimator, but rather to have a payment mechanism that reflects the way in which the contract price was developed and the work to be done on site.

Under the provisions of *The New Engineering Contract*★ estimators will be encouraged to adopt activity schedules rather than bills of quantities. It is anticipated that their use will help to reduce the likelihood of disputes and time overruns by promoting a close relationship between time and money which stimulates good management practice.

The estimator's prediction of the level and type of resources needed, made before the job starts, is given to the client, thereby placing the estimator in a strong position to claim for truly unforeseen conditions and valid delays and disruptions. If the contractor then underresources the job he will get little sympathy from the Engineer if progress is slow.

The issue of whether or not the client should have an open copy of the tender make-up is currently contentious. If he does have a copy and there is a change in the use of resources there can be no dispute about what the original intentions were, but this is commercial information and should perhaps be produced only in certain circumstances. One solution is to supply the make-up in a sealed envelope which is opened only in the event of a major claim or variation.

Some clients now require the submission of resourced programmes of work as they can see that this strengthens their position when evaluating change. The stage at which this is submitted is important. If it is at tender stage, there will have to be a high degree of collaboration between the estimator and the intended project manager so that the job is resourced realistically. This raises questions as to how much time project managers can afford to devote to projects that they are not certain of being awarded. If the resourced programme is produced after the award of the contract the contractor may produce a programme to match his strong position, and not one that reflects the tender. Nevertheless, the tendency towards client and contractor using common information can only produce greater clarity in communication and greater equity in decision-making.

Information technology is a key factor in the development of estimating practice because of the continuous developments in hardware and software which offer estimators greater speed,

---

★ Institution of Civil Engineers. *The New Engineering Contract.* Thomas Telford, London, 1993.

increased data handling capacity, more user-friendly and interrelated systems—often at reduced prices. The current trend seems to be to move away from using information technology for data capture, manipulation and retrieval in massive databases and towards its use in less mathematically rigorous methods of analysis for tackling estimating problems which may contain uncertainty. The future areas for development are more likely to be found in the newer generation of computers and quantitative methods of analysis, like fuzzy logic, which might be combined with a parametric approach to produce earlier and more accurate estimates for projects.

In conclusion, many of the problems caused for managers of construction projects might be eased by greater effort in the preparation and use of estimates. Only with this increased effort will it be possible to distinguish between poor project performance and a poor project cost estimate.

Biographies of the authors

EurIng **A. N. Baldwin**, *BSc, MSc, PhD, CEng, MICE, MCIOB*
Andrew Baldwin is Senior Lecturer in construction management in the Department of Civil and Building Engineering at the University of Technology, Loughborough. He has a background in civil engineering and has worked with several major contractors and a firm of leading consulting engineers. He has several years' experience in the development and implementation of computer-based construction management systems within several different companies including Trafalgar House Construction. He is co-author of *Estimating and tendering for civil engineering works*, and is the principal author of two publications for the International Labour Organisation in Geneva. His research interests are centred on the use of information technology within the construction industry and on how these new technologies will change current business processes.

D. Bower, *BEng, AMICE*
Denise Bower is a Lecturer in the Project Management Group at the University of Manchester Institute of Science and Technology. She graduated in civil engineering, worked with a contractor on site, and then moved into her current research post working on contract payment systems in civil engineering. Her recent work includes evaluating computer software for project management and earned value techniques for Ministry of Defence contracts, case studies of the management of major projects on an operating production site, the assessment of a process facility built under a concession contract, and recommendations of contract strategies for overseas projects. She is the joint author of several case studies of the management of recent construction projects.

R. Fox, *BSc, CEng, MICE*
Richard Fox has had extensive experience of fixed price cost estimating for process plant projects. He has gained particular experience of the problems caused by the conflicting objectives of the client and the contractor during the early stages of a project. Richard is currently Estimating Manager for Costain Engineering and Construction, Process Contracting Division.

P. Jobling, *BSc, MSc, CEng, MICE, MAPM*
Paul Jobling specializes in the management of construction projects. His experience includes work for contractors and consultants in the UK and overseas on large oil and chemical plants and major transport infrastructure. He was a senior member of Eurotunnel's project management team for the Channel Tunnel. He has also carried out research into cost estimating for high-risk construction projects and has published widely in the fields of cost estimation, project control and risk management. He is a principal consultant with Mouchel Management Ltd.

D. J. Lowe, *BSc, MSc, FRICS, CertEd*
David Lowe is a Lecturer in the Department of Building Engineering at the University of Manchester Institute of Science and Technology. After graduating from Trent Polytechnic in 1981, he gained experience as a quantity surveyor on building and civil engineering projects before becoming a Lecturer in 1986. He is currently undertaking doctoral research into the effects of experiential learning on early stage estimating.

EurIng **R. McCaffer,** *BSc, PhD, FEng, FICE, FCIOB, MBIM*
Ronald McCaffer is Professor of Construction Management, Head of Civil Engineering, and Dean of Engineering at Loughborough University of Technology. He is a Project Director of the European Construction Institute. After graduating in civil engineering at Strathclyde University, he gained industrial experience with Babtie, Shaw and Morton, the Nuclear Power Group and Taylor Woodrow before going to Loughborough. He is co-author of *Modern construction management*, *Worked examples in construction management*, *Estimating and tendering for civil engineering works* and *Managing construction equipment.*

P. Medley, *MSc, CEng, FICE, FIWEM*
Paul Medley graduated from Manchester University in civil engineering in the late 1950s. He then spent five years as an engineering assistant with the sewage disposal department of Sheffield Corporation. He then spent two years as a site engineer with Taylor Woodrow Construction where he worked on several major projects. He joined the water industry in 1966 and on its re-organization in 1974 became Divisional Engineer of the Derwent Division of the Severn Trent Water Authority. In 1983 Paul Medley transferred to the Upper Trent Division and in 1987 he took up his present post as Engineering Manager of the Burton Office of Severn Trent Engineering where he is responsible for the design and construction of water supply and sewage disposal projects.

M. Skitmore, *MSc, PhD, FRICS, MCIOB*
Martin Skitmore is Professor of Quantity Surveying and Construction Economics in the Department of Surveying, University of Salford. His research interests include estimating, bidding, cost modelling and the application of information technology to economic decision-making. He is author of many papers and the books *Contract bidding in construction: strategic management and modelling* (Longman) and *The accuracy of construction price forecasts: a study of quantity surveyors' performance in early stage estimating* (Salford University). He is a past chairman of the Association of Researchers in Construction Management and a member of the International Council for Building Research, Studies and Documentation Working Commission W-55 and W-65.

N. J. Smith, *BSc, MSc, PhD, CEng, MICE, MAPM*
Nigel Smith is Senior Lecturer in the Project Management Group, Department of Civil and Structural Engineering, University of Manchester Institute of Science and Technology. He is a Project Director of the European Construction Institute. After graduating in civil engineering at Birmingham University, he gained industrial experience with Wimpey, the North East Road Construction Unit and the Department of Transport. In 1991 he was Chairman of a joint meeting of the International Association of Cost Engineers and the International Association of Project Managers on estimating, held in Trondheim. He is author of many publications

and co-author of the book *Guide to the preparation and evaluation of build–own–operate–transfer (BOOT) project tenders*.

A. Tyler, *MRICS*
Alan Tyler practised for 24 years in the construction industry before joining the Department of Civil Engineering at Loughborough University of Technology in 1984. His professional experience covers all aspects of quantity surveying on building, civil engineering and overseas contracts. His overseas work included a two-year posting as Project Manager in Amsterdam. He is currently the Course Director for the BSc course in commercial management and quantity surveying, having previously been Course Tutor for the MSc course in construction management. His teaching and research concentrate on contract procedure, design estimating and cost control, project management and quality assurance.

G. Williams, *BSc, CEng, MICE*
Gareth Williams has held senior management positions with engineering contractors involved in the design and construction of petrochemical and process plant. He has over 30 years of experience on projects ranging from primary and secondary pharmaceutical manufacturing plants to large oil and chemical installations. He has a wide knowledge of various types of contract and the pitfalls likely to be encountered in the bidding and execution of industrial plant. He is currently working as a freelance consultant.